AF477391

Innovation and Nanotechnology

To Amelia

Innovation and Nanotechnology

Converging Technologies and the End of Intellectual Property

DAVID KOEPSELL

BLOOMSBURY ACADEMIC

First published in 2011 by

Bloomsbury Academic
an imprint of Bloomsbury Publishing Plc
36 Soho Square, London W1D 3QY, UK
and
175 Fifth Avenue, New York, NY 10010, USA
Copyright © David Koepsell 2011.

This work is published subject to a Creative Commons Attribution
Non-Commercial Licence. You may share this work for non-commercial
purposes only, provided you give attribution to the copyright holder
and the publisher. For permission to publish commercial versions
please contact Bloomsbury Academic.

CIP records for this book are available from the
British Library and the Library of Congress

ISBN 978-1-84966-343-4 (hardback)
ISBN 978-1-84966-480-6 (ebook)

This book is produced using paper that is made from wood grown
in managed, sustainable forests. It is natural, renewable and
recyclable. The logging and manufacturing processes
conform to the environmental regulations of
the country of origin.

Printed and bound in Great Britain by
the MPG Books Group, Bodmin, Cornwall

Cover image © kentoh/Shutterstock.com

www.bloomsburyacademic.com

Contents

Foreword

Sir Harry Kroto

As a child I had quite a lot of interests and in adolescence consciously tried to be as good as possible at several things, including drawing, tennis, and building things out of Meccano in the front room of our house which was my world as well as my school work. I did not really care about being the best – just about being as good as I could be. Meccano I think was quite important and I have written and spoken about it on several occasions as I think it enabled me to develop good manual dexterity, an understanding of engineering structures, and a feeling for the intrinsic differences between various materials from steel to aluminum and plastic. After graduating with degrees in chemistry and becoming a scientist in a university where I carried out research and taught students, my professional work mixed science and technology and also I did a fair amount of graphic design in whatever spare time I had. I also played a lot of tennis. Looking back I think the wide range of interests was a major factor in the cross-disciplinary approach that my research followed. It enabled me – essentially unconsciously – to find ways in which my interests overlapped with those of some of my colleagues and led to key breakthroughs in phosphorus/carbon chemistry and the properties of long linear carbon chains. These early studies led directly to discoveries of new and unexpected large carbon chain molecules in the interstellar medium and ultimately to the discovery of Buckminsterfullerene, C_{60} (buckyballs), and recently to its discovery in space.

Science seeks to uncover the laws of nature, whereas technology applies that knowledge. However, the two domains are inextricably linked as our experience has shown that new scientific discoveries invariably lead to totally unexpected new applications. The body of scientific knowledge painstakingly assembled by scientists and technologists forms a massive cache upon which successive generations can build. Sometimes, along the way, a new and unexpected breakthrough occurs of sufficient significance that a technological revolution occurs. Most scientists who make major breakthroughs acknowledge the debt their discovery owes to the work of others.

Major breakthroughs invariably depend upon collective action, the free accessibility of prior knowledge, and unselfish adherence to the ethos of science. Unfortunately personal desires for recognition and perhaps financial gain can easily interfere with progress. Many would have us believe that competition is the mother of invention. Today, many deem it a virtue, and the myth that cut-throat competition is needed for social and technical improvement is now deeply ingrained in our research culture. Certainly, lone geniuses and doggedly hard-working craftsmen driven by the desire to become wealthy have devised groundbreaking improvements, but as with all major advances, the breakthrough could not have been made without free knowledge of previous work. Without all that came before, and unfettered access to the growing body of general knowledge about the workings of the universe, progress would come to a screeching halt.

Interestingly, our molecule, C_{60} Buckminsterfullerene, has become an iconic cross-disciplinary symbol of the way scientific – in particular chemical – concepts and advances can generate interest and ideas in several other areas from engineering to the arts. It is not just an elegant, highly symmetric form of carbon with many potential possible applications. It is also a structure that captures the imagination of many people from professional architects to very young children. Graphite and diamond were previously the only well-characterized forms of carbon, and the fullerenes and their elongated cousins, the nanotubes, as well as graphene, promise to revolutionize materials sciences. They promise paradigm-shifting applications in the future if we can overcome some rather tricky technological problems. Their most interesting promise lies in their applications in nanotechnology. Research in the field of fullerene science is resulting in approximately 1,000 papers each year as researchers around the world uncover new properties and devise new applications. These new properties may well form the basis of an entirely new field of manufacturing, and the eventual realization of true nanoscale manufacturing, that is bottom-up assembly of the next generation of complex devices with advanced function.

Although there were claims of prior knowledge relating to the discovery of C_{60}, they had no real credibility and certainly no intellectual validity. Thus scientists have been free to experiment with fullerenes without the impediment of restrictive patent issues which often hold researchers, especially in industry, to ransom. I personally did not get involved with any patents on either the creation of C_{60} or the molecule itself nor had I any interest in such. I have no doubt that attempts to monopolize the use of the knowledge we have gained about how the natural and physical worlds work only stand in the way of progress.

It is a myth that competition is necessary for progress and we must find a better way to encourage young people to explore the way Nature works and use any knowledge gained only for the benefit of society. David Koepsell in this book explores the possibility that the emergence of nanotechnology will overturn previous ideas about the nature of all technical artifacts. He suggests that the ethos of science, by which each new discovery serves as the basis for the next, aided and abetted by intellectual *openness*, can be a more effective catalyst of technological advance. He argues that the institutions of intellectual property law are not just flawed and harmful, but illogical, unnecessary, and ultimately an impediment to innovation. The convergence of technologies embodied by nanotechnology, in what he calls *nanowares* (which encompass a range of technologies that are decentralizing modes of production), reveals the flaws. While he attempts to form a theory of artifacts based on first principles, he also examines the practical ways in which innovators in *nanowares* are adopting open methods of innovation, and are avoiding the pitfalls engendered by intellectual property issues.

Nanotechnology has a very long way to go before the paradigm-shifting technologies inherent in the properties of materials like fullerenes can be implemented. However, as described in this book, there are numerous grassroots approaches, as well as foundational work in the underlying sciences, that are paving the way. Even if Koepsell's notion of the death of IP, as revealed through our technology, seems premature, it is an important argument that we should consider carefully and recognize how legal issues are often part of the domain of discovery as well as invention. Ensuring that productive technologies emerge from basic science and enter the marketplace smoothly requires incentives to be carefully balanced in ways that institutions have often failed to achieve. If, as he argues, institutional strategies are so fundamentally flawed, and their collapse in the wake of *nanowares* is imminent, then researchers and innovators alike should look carefully at alternative approaches as he proposes. He argues that his approach, if adopted, will encourage basic science more effectively and lead to wondrous new technologies.

Sir Harry Kroto was joint winner of the Nobel Prize for chemistry, 1996.

Preface

I began exploring the nature of intellectual property (IP) while working on my PhD in philosophy and finalizing my law degree in 1995. The result was my first book, *The Ontology of Cyberspace: Law, Philosophy, and the Future of Intellectual Property*, and in it I called for the creation of a single, unified IP regime modeled upon copyright, but with shorter terms of protection. I believed that information and communication technologies (ICT) revealed that the old dichotomy between patent and copyright was unfounded, and suggested that copyrights were cheaper, involved no significant governmental involvement, and would suffice for protecting software.

After receiving my PhD, I practiced law, taught in a law school, worked in a software company, headed an international not-for-profit, did a post-doctoral fellowship at Yale University, and finally returned to teaching full time. Throughout these adventures, I have maintained an interest in IP law, its theoretical underpinnings, and its relation to innovation. Along the way I wrote another book, *Who Owns You? The Corporate Gold Rush to Patent Your Genes*, which explored ethical and ontological arguments against patenting unmodified genes. Because of that book, I learned that public philosophy has a role to play in developing institutions, and in public policy debate. Just a few months after the release of *Who Owns You?*, the Public Patent Foundation and the ACLU spearheaded a lawsuit against Myriad, a Utah-based corporation that owns patents on the 'breast cancer genes' (BRCA 1 and 2), or rather mutations to two genes in all humans that, when present, indicate a significantly increased likelihood of getting breast cancer. Many of the same arguments I had made in my book regarding the injustice of obtaining patents on naturally occurring, unmodified genes, and the pernicious effects on innovation caused by such patents, were at the heart of the lawsuit. As of this date, the plaintiffs have won in the district court, where Judge Sweet found that the patents on those genes are invalid attempts to monopolize natural products. That case will doubtless work its way up eventually to the Supreme Court of the United States.[1] The lawsuit affects millions of women (and men), and illustrates that issues of patent eligibility are not mere metaphysical ponderings. Women whose health insurance does not cover such diagnostic tests are now forced to pay more than US$3000.00 for a test because Myriad has a monopoly. It was with that suit, and upon meeting people directly affected and involved by this fight, that I became aware of the gravity of a patent system gone out of control. I also learned that, when cornered, the patent industry will lash out and dig in.

The 'patent industry' is what I call the entrenched interests not only of corporations and individuals who profit by the state-sanctioned, artificial monopolies of patent and copyright, but also the tens of thousands of patent professionals, bureaucrats, and their employees engaged daily in the patent system. They have tremendous resources, lobby groups, PR campaigns, and political influence to ensure their continuation, and if possible, the extension of their domain. Every patent that is filed is profitable … to the patent attorneys who do the filing. Meanwhile, somewhere between 2 and 6 percent of patents earn their costs back and make a profit for their inventors. The rest are worthless, or even arguably a drag on innovation. The patent industry naturally lashed out at the ACLU, the Public Patent Foundation, and activists who voiced their support for the case against Myriad.

The past 100 years have seen the growth of this industry, and any threat to its dominance will doubtless be attacked. I too was a subject of their ire, and even while philosophers, activists, the mainstream press, and even a handful of attorneys embraced the arguments I made in *Who Owns You?*, negative reviews came almost exclusively from patent professionals, except for the iconoclastic Stephan Kinsella, who is a practicing patent attorney who fully understands the problems we will delve into more fully in this book. I expect these patent professionals will be similarly uninspired by the scenarios I will paint here, and they will likely feel threatened by the future I predict.

Both of my past books included calls for action, for public policy change, and for modification of current IP schemes. This book will not. Rather, here I will explain why the ultimate demise of IP is *inevitable*, why the technology I call "nanowares" makes that so, and what innovators can do to prepare for it. I will discuss some of the ethical implications involved with nanowares, both within and outside the institutions of IP law, the economic consequences of its demise, and what I believe that nature of our relationships to artifacts really is. But as opposed to my past work, this is an attempt at more or less purely descriptive metaphysics, with some practical advice as to how to handle a tricky, transitional phase as institutions evolve and begin to better reflect reality.

Along the way, I will address some of the more commonly occurring concerns in nanowares, including potential risks, security concerns, and what duties scientists and innovators might owe to the public. But primarily, this book is an argument about the nature of types of expressions (artifacts), historically perceived needs to protect those artifacts through legal institutions, and what those institutions imply. I will include some brief case studies, to put into perspective the philosophical issues I am trying to elucidate, and to provide food for thought about how innovators and scientists can work to ensure that nanowares become a fully realized technology, and that their ultimate benefits are finally achieved.

This book is aimed primarily at those who are seeking to achieve the full potential of nanowares, either in foundational work in the underlying science and technology, trying to ultimately build molecular nanotechnology (MNT) components or systems, or those who are working at the grassroots to bring 'desktop' fabrication technologies to the masses. Self-replicating replicators, or cheap, home-made, and accurate three-dimensional printers which anyone can use to fabricate working prototypes of new things, will help to do for the real world of physical objects and innovation what the internet has done for innovators in video, music, and software. New markets can flourish, and market entry will be unimpeded by the need for capital that hinders innovation for all but the well-capitalized. This is the future I hope to see, and that is, as I will argue, inevitable. It is also the future that will completely and finally undo IP. People actually working in these fields know this is true, and they are already embracing institutions and approaches to the science and the technology to ensure that this future will occur.

This book is for them, and those who want to see them achieve a world without scarcity. I hope it provides some theoretical justification, and perhaps a bit of insight into the trends I will discuss, and why they are not just inevitable, but good.

David Koepsell
Leidschendam, The Netherlands, 10 September 2010

Acknowledgments

I am indebted as always to more people than I will recall or can name, though I will try my best to be complete. My wife Vanessa has encouraged and supported my work emotionally, intellectually, and physically, keeping me fed, keeping me thinking, and doing her best to keep me sane. I am eternally grateful to her. My colleagues at the Delft University of Technology have spurred my thinking with their questions, challenges, prodding, and support. I am especially thankful to Jeroen van den Hoven, Peter Kroes, Sabine Roeser, Ibo van de Poel, Behnam Taebi, Nicole Vincent, Floris Kreiken, Hu Mingyan, Maarten Fraansen, Pieter Vermass, and Rossitza Rousseva, all in no particular order, and among many others. Stephan Kinsella, David Levine, and Michele Boldrin have inspired much of this work, and I am thankful for their knowledge and e-mail correspondence. As always, my parents have provided emotional and editorial support as faithful proofreaders and honest critics. Thanks are due to Shane Wagman, who helped research alternative forms of protection for artists, and wrote up the brief case studies included about these alternatives. I am thankful to the Synth-ethics team, including Patricia Osseweijer, Julian Kinderlerer, Armin Grunwald, Laurens Landeweerd, John Weckert, Seamus Miller, Michael Selgelid, Elena Pariotti, Maria Assunta Piccinni, Christopher Coenen, and Hans-Jürgen Link. My dear dog Luna kept me company while I wrote, and walking her provided additional time for contemplation. And finally, my daughter, to whom this book is dedicated, whose imminence helped keep me on track, and whose promise inspires me to do what I can to make the world a better place as best I know how.

INTRODUCTION

Nanowares: Science Fiction Futures and Present Potentials

Consider this: everything that surrounds you is matter. Matter comprises every object in your sight, everything your body is composed of, processes, excretes, and thinks. Yes, every thought in your head is dependent upon matter as well, and as we know because of Albert Einstein, even energy is matter in another form. The atomic theory, which posited that the universe is composed of very tiny bits and pieces, each of which is further composed of tinier bits and pieces, is one of the most successful theories in science, having been confirmed by a hundred years of observation. The bits and pieces that compose the universe are the atoms, of various sizes and qualities, that we are now all familiar with from the Periodic Table of Elements. In the mid-twentieth century, as the atomic theory continued to be successfully confirmed, and humans grew increasingly capable of seeing and doing things with the world of the very, very small, some forward-thinking scientists began to wonder whether there were physical limits to the ability to manipulate atoms, and if not, what such limits might mean for our abilities to remake the world around us.

If, they thought, we could manipulate matter at the atomic level (at the 'nanoscale', a scale measuring one to one hundred nanometers, or one-billionth of a meter), then conceivably, we could develop very tiny machines that could in turn build things for us at both the micro- and macro-scales. Imagine if you could construct nanoscale robots (essentially, the size of large molecules) and instruct them to build a computer, or a car, or anything you desired, from readily available raw materials. Or consider the possibilities of using such nanoscale machines to combat diseases or parasites, viruses or cancers in the body, eliminating the need for the scattershot or blunderbuss approaches of much of modern medicine. Or perhaps we could develop radical new materials, incredibly strong, capable of mending themselves, light, efficient, or even imbued with properties we only dream of in science fiction. The possibilities of nanotechnology, which is the application to engineering artifacts of our increasing knowledge about physics at the nanoscale, are wide open.

The history of technology in general is a history fueled largely by the science of slow miniaturization and incrementally better tools for manipulating things at ever smaller scales. The past hundred years of electronics demonstrate this evolution, as vacuum tubes became transistors, and as silicon chips became quantum processors. The science of the small, from which the technology of computing has remarkably benefited, is now reaching out beyond the two-dimensional world of silicon wafers, and promises to remake the world of more-ordinary things, the hardware that composes the rest of our technology and human artifacts, and that may revolutionize the way we interact with our environment and tools. The history of computing serves as not only a catalyst for the coming age of nanotechnology, in which we will be able to begin to manipulate the world around us so that we can build remarkable new tools literally from their constituent atoms, but also as a suitable departure point for a discussion of the ways in which nanotech will alter the ways we innovate, and how we both encourage and protect innovation.

For about the past two hundred years, we have split the world into two. On the one side stood our artistic and creative artifacts, and on the other our utilitarian inventions. This dichotomy has been increasingly important, and lately challenged, as our artifacts became more closely tied to digital expressions and electronic media. Computerization, brought about largely through substantial breakthroughs in miniaturization, ushered in a new age of innovation in which machines and aesthetic expressions began to merge. The legal paradigm encompassed by intellectual property (IP) law, and embraced by a larger culture, began to break down.[1] Machines and aesthetic expressions no longer seemed so distinct. How can we learn from the mis-steps and impediments posed by the failing two-world paradigm, and build new institutions, both legal and cultural, that are not only philosophically sound, but that can encourage the sort of innovation promised by futuristic nanotechnology?

This book comprises an argument that (a) current schemes of IP protection are not only pragmatically incapable of being applied to nanotechnology,[2] but also theoretically inadequate to promote modern innovation, and (b) nanotechnology reveals that our relationships to our artifacts have been misunderstood, and poorly applied through laws that were once considered necessary to promote innovation, but which now prove to be only impediments. This argument is made by looking first at general trends in the history of innovation, and the technologies and institutions that have enabled and sped it in the past hundred years, converging now with nanotechnology in its broadest sense; and then looking at the first principles involved in our theories of IP, and criticizing their bases as well as effects. Finally, a new paradigm will be offered, based upon methods and processes being adapted in both the technologies of innovation and new forms of protection being developed at the grassroots, eschewing old forms of monopoly, and embracing both the spirit and methods of open science and innovation.

But first, we'll look at the general trends of industrialization, miniaturization, delocalized production, and of course, the 'convergence' that all of this is centering upon in modern and 'futuristic' nanotechnology.

From Arrowhead to Atom

Often, when we hear the term 'artifact' we think of ancient tools discovered in strata of ancient soil. But the term artifact simply means anything purposely created by humans in some enduring medium. Thus, a song is not an artifact when sung, since it drifts away into the aether never to be heard beyond its immediate performance, even though it is intentionally created by humans. However, an ancient recording of a song on a 78 rpm platter would be an artifact. A skeleton is not an artifact, until someone carves something decorative or useful upon it or arranges it into some purposeful position.[3]

Proto-human artifacts date back millions of years. The archeological record holds increasingly older surprises as we find that ancient humans have been turning found objects into tools at earlier dates than we even recently suspected. The first human artifacts came about the moment some early proto-human mixed labor with a found object with the intent to create something new. This is the genesis of craft, of art, of technology. Intention, as we shall see is critical, and is what makes artifacts different from accidents.

The history of human art slopes gradually upward from the first-fashioned arrowhead or similar tool, slowly tracing a halting curve toward the modern industrial age, becoming hyperbolic in the last fifty or so years. Most of the milestones upon that curve involved changes in the manner and means of production. The course of that history will be sketched here only in brief, and in broad strokes, because the way we evolved from chipping away at flint and bones, to being able to assemble machines from the atomic scale on up, illustrates the critical change that is posed by futuristic nanotechnology to our assumptions regarding our relations to our artifacts, and the ways we have come to encourage innovation through institutions.

Ancient technology was decentralized. Even while the manner of producing woven baskets, clay pots, ceramics, and eventually metal-ware demanded increasing skills and specialized knowledge, the artisans who made the crafts that formed our most ancient technologies created their goods by laborious, local, and more-or-less *ad hoc* manners. In other words, when an arrowhead was needed, it was fashioned most likely by the end-user, and perhaps in some surplus as necessary for an upcoming hunt. Perhaps surpluses were traded for goods produced by those with other skills and arts, and some form of ancient trade began to create specialties. But this sort of barter cannot be mistaken for a market, and technologies not otherwise impelled by more than mere necessity (no profit motive, for instance) grew only incrementally with the advent of new necessities. Thus, as ice ages waxed, new needs for clothing impelled the creation of new, local arts that could meet new needs and sustain life.[4]

This mode of production and dissemination of technology marks the first major epoch of technology. Driven by necessity, requiring only modest skills, and only very little in the way of specialization, the first human artifacts changed very little over the course of humanity's first million or so years. Up until about 10,000 years ago, in the Paleolithic or 'Stone' age, human tools were created as needed, from simple and available materials, and show only a modest degree of slowly increasing artistry. Human arts in general seem to have fulfilled the distinct purpose of survival for that first million years, aiding with the hunt, both materially and spiritually, as humans wandered from place to place. It was only with the invention of agriculture, and the advent of the Copper and Bronze Ages about 10,000 years ago, that human technology began to advance significantly, and the rate of change began to increase.

With agriculture came an end, more or less, to nomadic life. And with a more stable lifestyle, and the beginning of what we might call *culture* or even *civilization*, came the ability to specialize tasks more discretely, and to build more permanent means for creating artifacts or techniques. It is interesting to note that there are two very different and ancient means of creating an artifact, both of which emerge very early on, but one of which comes to surpass the other. Things can be made by either (a) taking something away from an existing thing (by chipping or carving bits away), or (b) 'building' a thing from some formless mass. For example, a piece of flint can be chipped away until an arrowhead remains, or a clay pot can be molded from formless clay into a thing with form and function. Most of the artifacts we are surrounded by today fall into the second category, and it is the paradigm behind many ideas for creating nanowares, by which anything could be 'built' from the bottom-up, and eventually atom-by-atom. These two basic means of producing artifacts will be discussed in more detail as we begin later on to delve a bit into the metaphysics of nanotechnology, but it's also worthy of note that the move from *a* to *b* was impelled and made possible by a shift in the necessities driving the creation of artifacts, and the modes and means of production that became available as humans began to settle into communities.

As nomads settled into stable groups, villages, towns, cities, etc., with cultures, so too did technique and artifacts begin to flourish. Without the challenge of everyday necessity impelling and permitting only that technology which affords immediate survival, artisans and craftspersons could commit more time, more resources, and more thought to developing tools meant to fulfill not just needs, but *interests*. Moreover, while the flexibility of crafts that rely upon found objects becoming fashioned into useful goods is low, new materials, primarily metals, pottery, and ceramics make building things from scratch a preferred and much more flexible mode of creation.

The techniques available for crafting new things out of flexible media (as we'll call, for now, clay, metal, and other media from which things can be built 'up', rather than cut 'down') could further be perfected as leisure time increased. As artisans became more adept at a particular, chosen technique, specialization became possible. Categories of craftspersons could thus improve in their practiced techniques as artisans need not be 'jacks of all trades'. All of which forms the basis for the development of an economy, as specialists learn to trade and barter their arts with others, producing their creations in strategic surpluses, affording themselves the option of trading their goods for those of others. Around 5,000 years ago, the first writing systems emerged out of the need to track trades of goods, and the wheel became adapted for carrying goods across distances, between various trade centers, expanding the scope of what we'd now call 'markets' for goods.[5] The emergence of trade and barter, the development of dispersed markets, and the luxury of specialization required certain manners of production, storage, and dissemination of goods. Surpluses became a strategic necessity, replacing the production of things as needed with a system by which one could begin to profit, and accumulate wealth, enabling the slow emergence of more modern economies.[6]

With specialization, crafts could develop into trades. Trades then developed their own cultures, policing their ranks, and ensuring market domination over a territory. This marks the next stage of the development of human technologies and markets, as arts became specialized, and the first, nascent form of 'IP' protection began to emerge. Secret-keeping, enforced through trade associations and guilds, is the first means by which those who practiced useful arts sought to control their markets. As technologies became complex enough to require training to produce or practice, and too difficult or time consuming to easily 'reverse-engineer', secret-keeping became an effective means to ensure that trades could dictate higher prices for their goods by policing the market and preventing undue competition. This in turn propelled the continued perfection of technologies, to prevent reverse-engineering, and to help ensure market domination by a particular trade.[7] Specialization and trades came to dominate the emerging market for artifacts for much of the past two to four millennia. But the growth of a new manner of acquiring knowledge, beyond the scholastic method (in which masters taught apprentices, and knowledge was kept obscure and secret from the uneducated masses), began to succeed where scholasticism failed, and thus began to assume primacy as humankind's foremost method of inquiry. Science slowly emerged as the dominant epistemological paradigm in the West during the Renaissance and Enlightenment, and led Europe and the Americas into the industrial revolution beginning in the late 1700s.

Science demands unfettered inquiry into the workings of nature, and replaces the confidence previously demanded over rote knowledge with a practiced skepticism, and ongoing investigation. With the rise of the age of science came the need to develop new means of treating information. Scientific investigations conducted by 'natural philosophers' could only be conducted in full view, out in the open, with results published in meetings of scientific societies and their journals. Supplanting secret-keeping and obscurantism, the full sunlight of public and peer scrutiny could begin to continually cleanse false assumptions and beliefs, and help to perfect theories about the workings of the world.[8] Science demanded disclosure, where trades and arts often encouraged secrets. And so as natural philosophers began to disseminate the results of their investigations into nature, new forms of trade, art, and industry began to emerge, as well as the demand for new means of protection in the absence of secrecy. Thus, as the scientific age was dawning, and helping to fuel a new technological revolution, *modern* forms of IP protection such as patents and copyrights emerged as states sought to encourage the development of the aesthetic and useful arts. By granting to authors and inventors a monopoly over the practice of their art, as long as they brought forth new and useful inventions (or for artistic works, as long as they were new), nation states helped to attract productive and inventive artisans and trades into their borders. These forms of state monopoly also enabled further centralization of trades and industries, as technologies now could become immune from the possibility of 'reverse-engineering' and competitors could be kept at bay by the force of law. This sort of state-sanctioned centralization and monopoly helped build the industrial revolution (by the account of many historians and economists, although this assumption has lately been challenged) as investors now could commodify new technologies free from the threat of direct competition, secure in the safe harbor of a state-supported monopoly over the practice of a useful art for a period of time.

In many ways, traditional IP was (and is) deemed vital to the development of large industries and their infrastructures, and to the centralized, assembly-line factory mode of production that dominated the twentieth century. With the benefit of a state-sanctioned monopoly, industry could build sufficient infrastructure to dominate a market with a new technology for the duration of a patent. This confidence assured investors that there would be some period of return on the investment in which other potential competitors are held at bay, at least from practicing the art as claimed in the patent. Factories could be built, supply chains developed, and a market captured and profited from, and prices will not be subject to the ruthless dictates of supply and demand. Rather, because of the luxury of a protected market during the period of protection, innovators can inflate prices to not only recoup the costs of investment, but also profit as handsomely as the captive market will allow.

For most of the twentieth century, IP allowed the concentration of industrial production into the familiar factory, assembly-line model. Even while the knowledge behind new innovation moved eventually into the public domain as patents lapsed, during the course of the term of patent protection, strictly monopolized manufacturing processes and their products could be heavily capitalized, and substantial profits realized, before a technique or technology lost its protection. But the modes and methods of manufacturing are now changing, and the necessity of infrastructural investment is also being altered by the emergence of new means of production, including what we'll call 'micromanufacturing', which is a transitional technology on the way to true MNT (molecular nanotechnology), and is included in our discussions of 'nanowares'. Essentially, assembly-lines and supply chains that supported the huge monopolistic market dominance models of the industrial revolution, well into the twentieth century, are becoming obsolete. If innovation and production can be linked together with modern and futuristic breakthroughs in micromanufacturing (in which small components can be fabricated and produced *en mass*, cheaply) and eventually molecular manufacturing (in which items are built on the spot, from the ground up, molecule by molecule), then we should consider whether the IP regimes that helped fuel the industrial revolution are still necessary, or even whether they were ever necessary at all. Do they promote new forms of innovation and production, or might they instead stifle potentially revolutionary changes in our manners of creation and distribution?

What we see when we look at the broad strokes of the history of artifacts, their innovation, production, and distribution, reveals clear trends. There is a general move from local, unspecialized, and decentralized to centralized and specialized, facilitated by institutions and states. There is a general move from use of available objects that can be fashioned into useful tools to developing tools from the 'ground up' from flexible media. There is a general move from the practice of arts in a scholastic mode to the incorporation of the methods of sciences into the process of innovation. There is also a general trend of miniaturization, facilitated most recently by the silicon revolution and computerization, which, as we'll see in the next section, parallels and predicts much of the coming issues this book is concerned with in the coming nanowares revolution, including prominently the role of IP regimes, and their effect on innovation.

A Very, Very Short History of the Very, Very Small

Precipitated first by the development of electricity, and the gradual move from a steam-powered world to an electrical one, miniaturization of components enabled both energy savings and smaller, more portable goods. While computing might have taken a mechanical turn, as the first calculators and rudimentary computers were mechanical (like Charles Babbage's Difference Engine), the success of new means for distributing electrical power achieved with Nikola Tesla's alternating current created a market demand for electrical goods, which in turn led to the development of vacuum tubes, then transistors, and ultimately silicon chips. Each successively smaller generation of products could be delivered to more customers, at more affordable prices, with less energy demand per unit, making operating costs also more affordable.[9]

Miniaturization began, of course, before electricity, primarily with timepieces used for navigation. John Harrison's famous H-series chronometers were built to meet a challenge posed by the British Crown, that of solving the biggest and most dangerous problem of naval navigation: longitude. Determining latitude was relatively simple given the angle can be determined from the pole by measuring the angle of relative declination of the pole star. But determining longitude was a singularly troublesome matter without accurate timepieces. While relatively accurate timepieces existed by the early 1700s, none were portable for the purposes of navigation at sea, as the shifting caused by the waves disturbed the mechanisms sufficiently to cause the best timepieces that existed to gain or lose time too greatly to be effective for long trips. Harrison solved this problem by developing very small, precise, spring-driven mechanisms for his timepieces. Miniaturization of chronographs, and their development into necessary fashion accessories, pushed the limits of mechanical miniaturization to their limits.[10]

In fact, clockwork mechanical computers were used for complex problems well into the Second World War. The larger these machines grew, the more power necessary to operate them, and thus steam, and electricity-powered devices were envisioned and built that could do rapid, complex calculations. To provide for storage and retrieval of information so that a general-purpose computer could run various different programs, punch cards and other mechanical means of read-only memory were incorporated into the earliest computers. The Jacquard loom is essentially such a computer, capable of reading cards with a weaving sequence programmed on them, and weaving the sequence automatically as the loom is powered. The electrically powered Mallock machine from the 1930s is an example of a mechanical, analog computer.[11]

The power and precision necessary for complex mechanical computers (especially analog ones – involving ranges of values *between* 0 and 1) helped push the development of digital computing and the adoption of logic gates in the form of vacuum tubes. Whereas reprogramming a mechanical computer could be accomplished through the introduction either of new mechanical elements or through punch cards, a truly flexible, random access machine, or 'universal computing machine' of the type theorized by Alan Turing, was made possible with digital, electronic computing. The machinery of a digital, electronic computer is its memory, in which both the data and software for manipulating that data can be updated without reconstructing the machine itself. The speed, flexibility, and eventually the size constraints posed by mechanical computation could be overcome with the introduction first of vacuum tube logic gates, storing data as '1s and 0s' (really, differing voltage states representing the numerals 1 and 0), and then as components became smaller and smaller, in transistors, and finally, silicon chips.[12]

Driving the need for ever smaller components were two factors: energy consumption and portability. The first vacuum-tube digital computers were monstrous, room-filling beasts that pumped out so much heat that they required enormous cooling mechanisms, and the combination of energy used by the computer and the air-conditioning made them prohibitively expensive to operate. Only wealthy universities, government facilities, and corporations could afford to own a computer and to operate one regularly. These sorts of computers were sufficient for the mostly military needs they fulfilled during the Second World War, but after the war, scientific and corporate demand for better, cheaper, and faster computing created a market for smaller, more efficient machines.[13]

The move to transistors and away from vacuum tubes as the means of computerized logic allowed for smaller, cooler, and more portable machines. The miniaturization of electronics helped fuel the consumer electronics revolution that drove an increasingly large amount of the western (and Japanese) technological economy through the 1950s, 1960s, and early 1970s. Transistors still relied on conductors that conducted heat poorly, and that impeded current, resulting in energy consumption and unreliability. Only with the development of the integrated circuit, and logic boards whose primary medium was silicon, could today's fast, cool, and relatively cheap computers be designed.

Energy efficiency and portability drove miniaturization in computing, and within a hundred years, the wildest dreams of Charles Babbage and others who were designing steam-driven, mechanical computers were exceeded by ubiquitous devices like modern cell phones, whose computing power is hundreds of times more than the best personal computers just twenty years ago. Silicon proved to be a magical medium, allowing the design and production, through photographic/chemical etching, of intricate and ridiculously small binary churning integrated circuits. The chips that are designed today have components so small you may need an electron microscope to see their full workings.[14]

Silicon has allowed for computerization to bloom, meaning that the range of human creativity that can be carried out by inputting data, manipulating logic gates, and outputting information has unleashed a torrent of new types of creativity and innovation. Take, for example, written expressions. At one time, shortly after the initial invention of the printing press, the ability to spread written communications was greatly increased, even while the producers of printed texts maintained a fairly secure monopoly over their trade through secrecy (printing presses were initially difficult and expensive to reverse-engineer, much less mass produce), and through intimidation as printers guilds developed. Printing technologies remained expensive well into the twentieth century, even when the technique behind the technology was well-known. The costs associated with acquiring a press, and the time and energy that were required to create original press runs of a book, newspaper, or pamphlets, kept the technology in the hands of those who could afford a significant capital investment. Now, because of the ubiquity of computers, and their ever-decreasing costs, creating printed goods is not much more expensive than the costs of the paper and binding. Anyone can, in effect, become a publisher and even, thanks to several small startups specializing in 'printing on demand', deliver bound, professionally printed books to anyone willing to pay the prices of shipping all over the world. If one forgoes the printed text, then publishers can instantly distribute their written expressions almost without cost anywhere that has internet access.

Or consider movies. As with printing, the tools of production were once prohibitively expensive for all but those with healthy amounts of capital to produce and distribute films. The cameras, film stock, and developing the negatives were all quite expensive, at least for professional quality productions worthy of wide distribution, and the infrastructure of film production and distribution were also controlled by trades with significant organizational inertia and power. But again, silicon and modern digital computers have liberated the means of both production and distribution, empowering the movement from idea to expression to become both rather simple and cheap, available to anyone with the creative wherewithal and just a little up front money. The primary requirement for realizing one's creativity is now *intellectual capital*. The same story applies to music, as well as to software itself, the engine behind the computer revolution.

These stories illustrate the trends that are broadly painted above: the move from art to trade, from delocalized to centralized, from cheap to expensive, and then back again. All of the modern computer and internet revolutions are built upon the move toward smaller, cheaper, and more energy-efficient modes of computing. Advances that have created modern computing are now being turned toward manipulating more than just silicon. These general trends have come, in a sense, full circle with *aesthetic* expressions of a certain type: those that can be realized digitally, with silicon-based computers and networks. The same general trends will mark our transition from the current mode of creating *objects* to the future or nanowares. When objects become as easily programmable and cheap to make as silicon-based expressions are, then anyone with an idea will be able to realize it in matter as easily as today's innovators in home-made texts, films, and songs. We should take heed then of the role of IP regimes in the computer revolution, and consider what we might learn from IP's role in the computer revolution to ease our transition to the nano-future.

Intellectual Property: A Concise History in Computing Technologies

Software has posed a problem for legal scholars and jurists since its inception, at least in the framework of IP that we have lived with for the last several hundred years. When IP began to make its debut a couple hundred years ago with statutes or acts of sovereigns in Europe giving publishers or inventors monopolies over their works, the distinction between a book and a printing press was clear. Books were passive machines for storing information, whereas a printing press was an active machine for making books. No one could have conceived that one day these two seemingly very different types of artifacts would converge into one powerful new technology. A computer fits neatly into our preconceptions of what a machine must look like, especially when computers still consisted of glowing tubes, blinking lights, punch cards, and other clearly machine-like appurtenances. Software, however, is another story. Software was not originally very different seeming than other parts of a computer. In fact, consider a computer that is programmed only for one task, like multiplication. In such a machine, the software is hard-coded into its logic gates, and those logic gates are clearly parts of a machine. The software is simply the hard-coded elements of a larger machine. Thought of this way, the software of any machine (even the design of a particular steam engine, for instance) is the arrangement of its specific parts in such a way that makes that machine perform its function as it was programmed (designed) to do. But the design of a machine is not software as it has *now evolved* to be. Re-programmable machines can perform new functions with the introduction of new instructions, and machines like these have existed before modern computers, pushing our definition of what constitutes a computer back further than we'd at first expect. We could consider inventions like the Jacquard loom to be early computers. These looms helped automate and standardize the process of weaving specific patterns into materials. When done by hand, weaving requires careful attention to when to throw the shuttle, and when to raise or lower the harness, producing a specified pattern. Human operators would read written notations which indicated when to do which action so that a pattern could be produced, and reproduced. Think of this written notation like sheet music. In 1801, Joseph Jacquard automated this process by fitting a loom to 'read' punch cards with holes that caused the raising or lowering to occur automatically, rather than requiring a human intermediary to read the notation. The machine reads it. The human-readable notation is substituted with a machine-readable software that eliminates a human step from the machine, speeding up the process, and helping to eliminate errors.[15]

The same thing happened a hundred years later with 'player' pianos, which anyone who has seen a Hollywood Western knows operates by 'reading' a roll of paper with punched-holes, eliminating the human intermediary called a piano player, and allowing the machine to read the music in software form. The technological evolution from Jacquard looms to player pianos is obvious, as is the move from hard-wired computers to punch card reading computers, but who knew that a legal case involving player pianos would make copyright unavailable to software until the 1970s as well as making it unavailable to vinyl records and audio tapes?

Expressions versus Machines

The rule used to be simple: in order to be an 'expression', and thus amenable to copyright protection, a writing had to be directly perceivable by a human. Words on a page, paintings, photographs, statues, and even sheet music would qualify under the 'direct perception' test. But player piano rolls were another story. In the *White-Smith Music Publishing Co. v. Apollo Co.* case of 1908, the Supreme Court of the United States made explicit the 'direct perception' rule, and thus excluded from copyright protection any expression that could not be directly perceived (or rather, understood, because humans can of course perceive holes).[16] As a result of this ruling, phonograph records, audio tapes, and other modes of recording information in ways that were not directly *understandable* by ordinary humans could not be protected by copyright. It took an amendment to the Copyright Act in 1972 to change this, and the direct perception test was finally abandoned.

The 1972 Copyright Act corrected a logical error in the law, which survives still in the dichotomy between patent and copyright, that distinguished between types of expression based upon human faculties. This error seems to have been based upon the notion that if humans cannot immediately understand an expression, it isn't really an expression. The illogic of this point of view is now quite clear. Expressions in foreign languages cannot be immediately understood by those who don't speak those languages. Some humans could likewise plausibly train themselves to understand mediated expressions like player piano rolls, just as they can learn to read sheet music. But sheet music, writing, the grooves on a vinyl LP, the patterns of 1s and 0s on computer tapes and discs are all media for transmitting ideas, and are all likewise expressions. The *purposes* of different types of expressions may differ, but if copyright law is meant to protect human expressions, then excluding a range of mediated expressions just because they generally require some further translation by another machine for their enjoyment makes no logical sense. The 1972 amended Copyright Act was doubtless impelled by pressure from music publishers and record producers to give copyright protection to their recordings, and to help ensure future profits from a growing base of music consumers, but the amendment had the likely unintended consequence of fixing the metaphysical absurdity of excluding machine-readable expressions from the broader set of *expressions* in fixed media.[17]

What is an expression anyway? How does an expression differ from a machine, and why does the law create the dichotomy between the subjects of patent and copyright? Before the US Congress fixed the artificial distinction among machine-readable and human-readable expressions in 1972, computer software was deemed patentable. The instructions in a computer, which open certain logical gates at certain times and close them at others, causing *computation* and outputting results, are *parts* of a machine. A hard-wired computer, as discussed above, can only do one function. But the same architecture could be re-wired to perform different functions, and modern computers are useful precisely because the instructions for opening and closing logical gates can be changed by the introduction of new software. The computer then becomes, technically, a new machine when the new software is input. It is one machine when it is off (one capable of performing any number of functions), another machine when it is turned on and the underlying software is activated (like an operating system), and another machine as new pieces of additional software are 'loaded'. In each case, the software that directs its actions is part of the total machine, making the computer an enormously (if not infinitely) reconfigurable machine, capable of taking on new forms. Which part of the machine is the expression? Should software be properly considered an expression if, as we can see, it is part of a machine?

Yes, software is an expression. The error of the law, corrected only partly still, is in distinguishing between a certain type of expression and a machine in the first place. At least with the amended Copyright Act, the artificial distinction between those expressions that are machine-readable or mediated versus those that are 'directly perceived' by humans has been knocked down. But in what sense is it proper to exclude machines from the category 'expressions' at all? A machine fails only to be a certain *type* of expression, just as punch cards failed to be a certain *type* of expression.

After all, a punch card in use is part of a machine, as is other software. But I have argued at length and for some time that the distinction among machines, books, songs, etc., is artificial, and not founded upon any sound metaphysics. In *The Ontology of Cyberspace*, I first argued that anything that is man-made and intentionally produced is an expression. Thus, the whole range of human artifacts, and even some less-than-clearly artifactual things (like a spoken word) are properly called expressions. The uses of these various types of expressions clearly differ. Some are meant to give us aesthetic pleasure, other expressions serve more 'utilitarian' ends. But the largest distinction underlying all of IP law, which says that machines are different than songs, is ontologically mistaken. It took the flexibility of computerized media to reveal this faulty ontology, and it still has not been corrected in the law, despite the many practical problems this has caused in the software industry.

The IP Mess in Software

The development of IP, once encouraged as a useful means of promoting innovation, creativity, and disclosure, may be doing the opposite in software. The first IP protection afforded software was patenting. But patenting is expensive to pursue, costing years of precious time before completion and often thousands of dollars. Copyright is free, you can do it yourself. Every author instantly owns the copyright of their own original works, even without affixing the © mark with their name and year of production to their work, if the work is 'fixed' in some tangible medium. Once the amended US Copyright act specifically included what ought to have never been excluded – expressions that were machine-readable – software authors began to copyright their works rather than patenting them. But this posed both a practical problem and a metaphysical conundrum, all of which pointed even more clearly to an underlying fault with IP law.

The realms of patents and copyrights were once held to be mutually exclusive categories. Something could never be both patentable and copyrightable. A wealth of cases and the well-established practices of patent offices stood for the proposition that a utilitarian object, as long as it was new, non-obvious, and useful, could *not* receive a copyright. Moreover, an aesthetically oriented 'work of authorship' could not be patented. Never mind that aesthetic pleasure is a use. I have often wondered whether the distinction is rooted in Calvinism, as the so-called 'useful arts' are rewarded with strong (though relatively brief) monopolies, and governmental intervention, whereas authors and artists are more or less left to their own devices for the protection and enforcement of their creative turf. But this strange distinction among legitimate uses resulted in a true conundrum with software, leading to a breakdown in previously mutually exclusive categories, and causing some to wonder whether software was some sort of new object that did not fit neatly into the previously clear categories of patent and copyright. I have argued that, in fact, these categories were poorly construed from the start, and that all expressions ought to be treated as a kind, and each new man-made object falls somewhere on a spectrum between 'mechanical' usefulness and other uses. Software has revealed for us the faulty ontology underlying all of IP law, and continues to wear away at its foundations.

Yet today, software patents are multiplying, seen as a favored apparatus especially for wealthier software companies to protect their innovation through strong monopolies, and to expand patent portfolios with which potential competitors can be excluded from the use of certain algorithms. Much of the value of software giants like Microsoft and Apple rests upon the size of their portfolio of patented algorithms. With these, and a cache of patent attorneys at their disposal, these companies can keep smaller competitors at bay. Patents are ideal for this given that there is no exception for 'honest' infringement. In other words, if you happen to develop a neat and potentially profitable algorithm, and never knew it was part of some previously patented code, you're out of luck. You cannot use it. You must either 'work around' it, or license it from the original patent holder in order to incorporate it into your invention. This is in stark contrast to copyright law which does allow for the independent, innocent creation of a substantially similar expression without infringement. This is perhaps why software patents have grown in popularity with those who can afford them, and why many smaller, newer companies are opposing software patents and working outside the system. Having seen the mess that software patents can cause, and the roadblocks to innovation that they feel are posed by patents on algorithms, many software developers are choosing either to rely once again on the ancient mode of protection known as trade secrets (Google's search engine is built on trade secrets) or using new forms of open development, including Open Source licensing, by which profits are gained not through an artificial, governmentally granted monopoly over an art, but rather through quick innovation, customer service, and moving good products swiftly to market.

Of course software is patentable subject matter, but then, if we are to be fair, so is James Joyce's *Ulysses*. When it was produced, it was new, non-obvious, and useful. It was an object, created by man, and a composition of matter, or even arguably a process. Under the current, broad interpretation of patentable subject matter, anything on earth made by man ought to be susceptible to patent. While some would argue that *Ulysses* lacks 'usefulness', this is an impossible position to defend for a work that has been found useful to so many consumers willing to pay to own it, to submit to the difficult task of reading it, and sometimes even enjoying it. The use is a primarily aesthetic one, but then so too are the uses of many machines at least partly aesthetic. And so while software patents are debated in the markets, in the courts, and in the media, none of these debates seem to get to the heart of the debate: why do we distinguish between types of expressions, when software has revealed for us the flaw in the distinction between patent and copyright? Why not have a single, unified scheme of IP protection that recognizes that all man-made objects intentionally produced (an important *caveat*, because some man-made objects are accidents) are expressions, and distinguishing amongst types of expressions is no longer meaningful?[18]

This has troubled me for the past decade, and the world seems no closer to resolving this problem, entrenched as we are in traditional forms of IP protection despite their growing obsolescence in the face of new technologies. If software didn't destroy the old IP paradigms, it has surely undermined them. Nanotechnology is poised to finally do what software failed to do: destroy IP for good.

Nanotechnology, Microfabrication, and the End of Scarcity

The trends broadly painted above illustrate a number of convergences, some of which will draw us into societal conflicts that may either propel innovation in a new direction or stymie growth and limit potentials. While much angst has been expressed about the safety and security risks posed by nanotechnology, these are more or less science fiction scenarios that do not concern everyday researchers in the field who are battling with the enormous physical and technical hurdles of developing anything like the futuristic nanotechnology in Michael Crichton's *Prey* and similar scenarios. Self-assembling nanobots are decades away, but the possibilities posed by the trends of delocalized fabrication, nanotech's precursor, could well be impeded by traditional IP practices and norms. The liberating possibilities posed by the ability to make nearly anything, anywhere, whether through current technologies (like Fab Labs or cheap, open source 3D printers, discussed more in subsequent chapters), will not be realized under current IP regimes, for many of the same reasons that software patents are impeding innovation in that marketplace. The challenge initially posed by the anti-innovative effects of software patents will leak into the world of 'physical' goods as all objects around us become more like software. This could be a tragedy, given the vast promise that new technological trends hold for solving all problems related to scarcity.

Scarcity is at the heart of the problem. The world of physical goods has long enjoyed robust markets, realizing profits for craftsmen and other producers, and abundant resources for consumers where those products were never afforded any IP protection. Grain, fruits, vegetables, the simplest farm implements, tables, chairs, knives, spoons, and nearly every other 'necessity' of life no longer enjoy any IP protection, and yet the markets for these types of goods survive. Supermarkets teem with the exchange of unprotected goods, whose value stems from traditional, well-understood mechanisms of supply and demand. Artificial scarcity in a truly competitive marketplace means that a competitor can undermine another's prices and defeat dishonest attempts to manipulate supply. Thus, smart farmers produce as much as the market will bear, and not much more. Warehousing perishable products won't work, and in a free market (though none really exist in agriculture as states create mechanisms to support local producers, thus defeating the forces of supply and demand to some extent) overabundance is no better than scarcity, as the price will surely plummet, and everyone will suffer. But markets for other types of goods are constantly manipulated, even as are those for agricultural products, not just by market players, but by governments seeking to provide advantages to their own producers, even where doing so may ultimately harm consumers.

IP is a form of artificial scarcity, created by governments to provide local producers with a market advantage that skews free markets. There is a very good reason proposed (but not proven) for this, at least from the viewpoint of encouraging innovation. If you don't think people will innovate unless scarcity can be maintained, then you create scarcity by granting a monopoly to the first to innovate in a particular art. The fear of 'freeloaders' profiting using the borrowed innovations of others is a real fear, but only if you think that what is being used by the newcomer should not remain freely available for use and profit. The theory behind the justice of IP laws depends upon accepting the ethical stance that the fruits of one's *intellectual* labor are somehow equivalent to the fruits of manual labor. For whereas we agree that if you build a house on a property you own, then you are entitled to occupy it unencumbered by anyone else's claims to your property, do we likewise necessarily agree that if you come up with a good idea and express it for the world to see or hear, then others cannot use it freely? Many argue that the two situations are not ethically equivalent, as in the latter case, you are deprived of nothing by the use of your idea by another, whereas in the former, a squatter taking up residence in your living room does actually deprive you of the exclusive use of a property you built, enjoy, and justly own. Again, it is scarcity here that defines the ethics of the ownership and exclusivity. Simply put, IP is quite unlike other forms of property.[19]

The lack of scarcity of IP in a world prior to IP laws was apparent to Thomas Jefferson, an early proponent of limited IP laws, and a principle author of the first Patent Act. I quote him here at length because his insights are particularly important today, as we move into a new age of innovation:

It has been pretended by some, (and in England especially,) that inventors have a natural and exclusive right to their inventions, and not merely for their own lives, but inheritable to their heirs. But while it is a moot question whether the origin of any kind of property is derived from nature at all, it would be singular to admit a natural and even an hereditary right to inventors. It is agreed by those who have seriously considered the subject, that no individual has, of natural right, a separate property in an acre of land, for instance.

By a universal law, indeed, whatever, whether fixed or movable, belongs to all men equally and in common, is the property for the moment of him who occupies it, but when he relinquishes the occupation, the property goes with it. Stable ownership is the gift of social law, and is given late in the progress of society. It would be curious then, if an idea, the fugitive fermentation of an individual brain, could, of natural right, be claimed in exclusive and stable property.

> If nature has made any one thing less susceptible than all others of exclusive property, it is the action of the thinking power called an idea, which an individual may exclusively possess as long as he keeps it to himself; but the moment it is divulged, it forces itself into the possession of every one, and the receiver cannot dispossess himself of it. Its peculiar character, too, is that no one possesses the less, because every other possesses the whole of it. He who receives an idea from me, receives instruction himself without lessening mine; as he who lights his taper at mine, receives light without darkening me.

That ideas should freely spread from one to another over the globe, for the moral and mutual instruction of man, and improvement of his condition, seems to have been peculiarly and benevolently designed by nature, when she made them, like fire, expansible over all space, without lessening their density in any point, and like the air in which we breathe, move, and have our physical being, incapable of confinement or exclusive appropriation. Inventions then cannot, in nature, be a subject of property.

Society may give an exclusive right to the profits arising from them, as an encouragement to men to pursue ideas which may produce utility, but this may or may not be done, according to the will and convenience of the society, without claim or complaint from anybody. Accordingly, it is a fact, as far as I am informed, that England was, until we copied her, the only country on earth which ever, by a general law, gave a legal right to the exclusive use of an idea. In some other countries it is sometimes done, in a great case, and by a special and personal act, but, generally speaking, other nations have thought that these monopolies produce more embarrassment than advantage to society; and it may be observed that the nations which refuse monopolies of invention, are as fruitful as England in new and useful devices. (Thomas Jefferson, letter to Isaac McPherson, 13 August 1813)

Jefferson's insights are repeated today in the modern mantra that 'information wants to be free', something adopted more or less as a motto by the 'free software' movement and other opponents of IP. The problem we are now faced with, at the advent of the nanotech age, is whether IP poses more of a threat to the elimination of scarcity, increasingly rapid rates of innovation, and generally magnificent promises inherent in the transition from 'dumb' to programmable matter.

You Say You Want a Revolution?

In this book, I attempt to close an arc I began tracing more than a decade ago when I considered the problem posed by software patents, and proposed, among other things, a way to solve it. In *The Ontology of Cyberspace* I argued that the underlying ontology of expressions suggested that we could create a unified scheme of IP protection, and eliminate patents entirely. In *Who Owns You?* I expanded upon my thinking about IP as applied to biotech products, specifically the strange case of gene patents, and suggested that there were parts of the world that were simply, ethically not prone to ownership claims. I argued that there exist not only commons that we create where property claims would otherwise be legitimate (as in state parks, or other lands that are set aside purposely for public use), or what I called Commons by Choice. There is another type of commons, a Commons by Necessity, that includes parts of the universe that are simply unencloseable either as a matter of practical necessity (like sunlight) or as a matter of logical necessity (like laws or products of nature – e.g. genes). Now I am concerned with a promising new technology that many consider to be the ultimate convergence of all technologies. Nanotechnology and its currently working precursors (discussed in greater depth soon) are delocalizing manufacturing, opening up the process of innovation, and liberating intellectual capital all over the world. These technologies, if fully realized, will remake our economy, help to end scarcity of material goods, and incidentally reduce harms to the environment created by the necessity for both surplus and storage, but also shipping of goods. I will argue that if we continue to embrace traditional IP regimes, we will ultimately hinder the full realization of the promise of nanotech.

A clash is inevitable, however, because those who stand to profit by the IP industry, and who cling to the use of artificial scarcity to ensure profits, are already threatened by a growing resistance to IP in current technologies. This is why we must now begin to choose which direction we will take, and how we can seize the ever more affordable tools of production, and work for institutional changes that will help us to truly liberate the full energy of human creativity.

Nina Paley, *Mimi & Eunice* (cc) creative commons

CHAPTER ONE

Let's Get Small (with Apologies to Steve Martin)[1]

From Feynman to Drexler

It is possible to see clear and broad trends in technological advances looking back over the past hundred or so years. Among these trends are ever-increasing efficiency in production processes and tools, the continued integration of computing and other technologies, the classical truth of 'Moore's law', which predicts the doubling of computing power every eighteen months, and miniaturization. In some ways, all of these trends are inter-related. In many ways, the challenges of the free market have driven all of these trends. When new technologies are introduced, they succeed or fail at the whim of consumers, and capital investments are gambled with each new technological roll-out. To increase profits and expand slim margins, efficiencies in the tools of production can hedge the bets of innovators without sacrificing potentially profitable new technologies. Computing has helped further expand margins in production by enabling robotics in manufacturing, and in helping to make better products with more capabilities. The hyperbolic climb of computing power only adds to these production efficiencies, making the tools of production increasingly smarter, faster, cheaper, and more energy efficient. Miniaturization adds to all of these efficiencies.

Some people seem quite prescient, and able to predict historical, economic, or technological trends with uncanny accuracy. Gordon E. Moore, who developed his famous law while working at Intel, the truth of which has been borne out by history, is one of these sages. Another is Richard Feynman, the Nobel prize-winning physicist. In a lecture he gave at the end of 1959 for the annual meeting of the American Physical Society at Caltech, he stated:

> I imagine experimental physicists must often look with envy at men like Kamerlingh Onnes, who discovered a field like low temperature, which seems to be bottomless and in which one can go down and down. Such a man is then a leader and has some temporary monopoly in a scientific adventure. Percy Bridgman, in designing a way to obtain higher pressures,

opened up another new field and was able to move into it and to lead us all along. The development of ever higher vacuum was a continuing development of the same kind.

I would like to describe a field, in which little has been done, but in which an enormous amount can be done in principle. This field is not quite the same as the others in that it will not tell us much of fundamental physics (in the sense of, 'What are the strange particles?') but it is more like solid-state physics in the sense that it might tell us much of great interest about the strange phenomena that occur in complex situations. Furthermore, a point that is most important is that it would have an enormous number of technical applications.

What I want to talk about is the problem of manipulating and controlling things on a small scale.[2]

Feynman then described futuristic, theoretical techniques now employed in electron microscopy to manipulate individual atoms, the benefits of storing large volumes of information at what we now call the 'nanoscale', the nature of biological machineries that are effectively nanosystems that 'do things' rather than simply store information, the potentials for miniaturizing computers, and some of the physical and technical challenges that would be faced before these breakthroughs could be achieved. It was a stunning moment in physics which all now recognize as the beginning of an era. Yet in many ways, it was also the necessary, incremental phase of something that had been going on in technology for more than a hundred years. All that Feynman did was coalesce previously existing and visible trends in technology, and predict their applicability and importance to future technologies. In the 1990s, Eric Drexler expanded on Feynman's vision, and gave further theoretical validity to the development of nanoscale manufacturing (the holy grail of nanotechnology, by which anything might be assembled atom-by-atom).[3] Vernor Vinge described the historical and inevitable convergence of technologies as the 'singularity'[4] and futurist Ray Kurzweil lays out a graph like that of Moore's law on which the general trend of converging technologies is superimposed, again with uncanny accuracy.[5]

Yet the trends in technology that Feynman and Kurzweil, and numerous others, correctly describe and predict are not trends in a vacuum. As with all human phenomena, all intentional ones at least, they are driven by human needs and desires. As such they follow the laws of economics, which is the science of predicting markets in light of evolving needs and desires. We should ask then not only what market forces drive the trends of technological convergence, but also what human needs and desires drive these forces? We might also consider what the effects of converging technologies on future markets will be, whether and to what degree those effects will be disruptive, and how we might adapt in ways that prevent the potential harms of significant disruption, socially, culturally, and economically.

Many have considered the potential harms posed by manufacturing at the nanoscale. Novels like Michael Crichton's *Prey*, and a famous essay by Sun co-founder Bill Joy, 'Why the Future Doesn't Need Us',[6] stoke both serious philosophical and ethical debate, and public fears. The potential for individual, physical harms from converging technologies is real, and there are already instances of nanoscale materials that have been developed and marketed, although they later turned out to be potentially harmful. But this is true of every new technology. Biotech faces similar potentials for harm and abuse. Even coal and steam technology will help alter our environment in potentially harmful ways, if man-made contributions to the greenhouse effect cannot be halted or reversed. While we should consider the potential harms and how institutions and principles might help prevent them, the inevitability and potentially revolutionary good that converging technologies pose argues that we instead seek to effectively capitalize on them, guided by our principles and concerns. We need not necessarily evoke, in a knee-jerk manner, the 'precautionary principle' that effectively set European investment and development of genetically modified foods and organisms back about a decade. Nor can we. The nature of the trends that Feynman noticed more than a half-century ago is that they are not only revolutionary in nature, but also inherently democratic. Like computer hobbyists who jump-started the modern PC revolution by bucking the IBM mainframe model and pursuing 'personal' computing in garages and basements, nanoscale manufacturing and its current precursors (like synthetic biology) are becoming accessible to those with modest investments in various tools.

The singularity is inevitable, and the only question that remains is what will we do to be prepared for it? We have some choices, made explicit through our prior attempts to deal with technological revolutions. We should consider our institutional and individual responses in light of the great potential for good posed by nanowares.

Technology Makes a Tiny Difference

Much of the literature and discussion of nanotechnology rests upon an assumption that this is a radically disruptive technology unlike others, and that the dangers it poses have never been faced before. Thus, many of these debates inevitably focus on harms and risks. While harms and risks are certainly something we should take seriously, they are by no means the entire story of the potential disruption posed by nanotechnology. Nor is it necessarily true that this disruption is unlike anything we have seen before. Let's put it in context. The context begins with the industrial revolution and ends in the nuclear age. At many stages of the development of technology, all the way from crossbow to H-Bomb, we can point to what in science has been christened 'paradigm shifts', involving tectonic changes in the ways we view the natural world.[7] Except in technology, these paradigm shifts mark changes in how we *interact* with the natural world, and in how we both develop and use *artifacts* (all man-made, concrete objects, intentionally produced).

At the beginning of the industrial age, the shift was away from individually produced artifacts, manufactured generally by individual craftsmen employing labor-intensive processes. With industrialization came the trend to employ labor in new ways, less for crafts and more for pure muscle. As steam power freed up time by speeding transportation, and freed up man-hours by devolving some labor to machines, the artisan class was replaced with a laboring class, and over time this laboring class developed both wealth and leisure time that encouraged the production of new goods for new markets. Industrialization marked a disruptive shift in the relation of people to goods, markets, and their individual labor. It was not without controversy, wringing of cultural-conservative hands, and violence. Luddism involved the actual destruction of the new machines by those who opposed the societal and economic changes brought about by industrialization. Marx decried the alienation of individuals from their own labor, and fomented revolutionary sentiment that changed international politics for a century. But the technology marched on.

Industrialization moved inevitably to mass-production, and the paradigmatic factory production line. The trend of distancing individual laborers from creativity, and using them as more or less mere operators of machines, continued through the twentieth century. Unionization helped increase the price of labor, and further encouraged the development of machines that could help replace expensive laborers. Mechanization required computation, and the trends in technology followed (or drove) trends in economics. The shift in the 1970s and 1980s from a manufacturing economy toward a service economy was as inevitable as the technological trends predicted by Feynman, Moore, Drexler, Vinge, and Kurzweil. But some of the potential effects of these trends were not well-predicted at all, and technological paradigm shifts have sometimes been unpredictably liberating, even as they were disruptive. Consider the computer revolution in the 1980s and 1990s. It was nothing like that predicted by those who had captured the computer market at the time. When IBM began selling business computers in the 1950s, it estimated a market for only fifty customers. It quickly had orders from seventy. Even in the 1970s, Ken Olsen, who was the co-founder of Digital Equipment Corp, said '[t]here is no reason for any individual to have a computer in his home'.[8] The same year saw Apple releasing its groundbreaking personal computer the Apple II, which had essentially been designed by Steve Jobs and Steve Wozniak in a garage. Who knew?

In retrospect, the path of the computer revolution was sewn into the fabric of the technology itself just as with each new disruptive technology. Technologies move toward consuming less power, and so they must become more efficient, and size matters for efficiency. Speed increases too with efficiency, and what once took a mainframe could be accomplished by ever smaller transistors, which became what we now call 'chips'. Mass-producing chips increased margins, and greater availability pushed down prices. All of these recapitulated trends in each new disruptive technology. 'Smaller, faster, and better' always leads to tools that eventually became more generally available. Moreover, a certain overarching human need or desire pushed computing to become a personal technology, and the PC captured a need that has never disappeared despite the advent of the industrial age: the desire to *create*.

Steam locomotives led to automobiles and motorcycles, mainframe computers led to desktop PCs, and the tools of production always tend to become cheaper, smaller, and easier to use. These trends are driving current precursors to nanotechnology which, when fully realized, will finally make everyone who wishes to design and produce new artifacts a potential factory owner, just as the PC has made publishing, film-making, and professional music recording accessible as never before conceived. In a very real way, the specific form of this new disruptive technology is not remarkable, nor are the potentials it offers. We have seen this before. But because nanotechnology will finally merge materials with programming, and authorship over the physical world will become possible, this particular paradigm shift will be felt at every level of society. How can we prepare for it, and still enable its full potential? Let's look briefly at the state of the art, and its real imminent and future potentials as well as risks.

Current Policy and Nanotech

Because much of the public debate and media attention paid so far to nanotechnology centers upon risks, there have been various national attempts to regulate the dangers of this technology by several governments. There have also been numerous scientific and public colloquia, conferences, and reports drafted regarding risks and regulations. Books too have been authored, ranging from a few monographs to dozens of collections of essays detailing the various ethical, social, and economic impacts of nanotechnology, and in some cases proposing manners of regulation and managing the coming revolution.[9] Meanwhile, in the United States there has been only sparse and sporadic public engagement and public policy initiatives to manage the transition toward converging technologies. Europe, and especially the United Kingdom following Prince Charles's well-known public panic about 'grey goo', has been more proactive. But in all instances, focus has been mostly on risks and harms, with little attention being paid to how to effectively manage the inevitable transition to a new mode of manufacturing, nor grappling with the social and economic consequences without significant upheaval. Questions that ought to be considered in more depth include: How can innovation be encouraged and profitable when matter becomes programmable? What will be the nature of authorship and inventor-status, and how can these statuses be protected? Should they be? To what extent can the tools of production be regulated when they will become ubiquitous, as computers have? Too little attention has been paid to these critical questions.

The regulations and discussions about risks are important as a frame for much of the future debate. While I am more optimistic about the promises afforded by nanotechnology, I am realistic about its risks. But realism means comparing risks with those of past technologies, and taking into account the reality of perceived versus actual risks over time. The regulatory climate so far has reacted realistically, but this should also imply that the scientific venture of delving into true risks proceeds the same way. Recently, concerns about the safety of certain nanomaterials have emerged from scientists' own research, proving that when they are not being manipulated by large corporations or other often conflicted forces, the institutions of science can discover risks and report them conscientiously. Specifically, nanotubes are a promising new area of materials research involving carbon structures designed at the nanoscale, which have potentially useful qualities like strength, flexibility, and conductivity. They also share some qualities, it seems, with asbestos. Like asbestos particles, which can burrow into tissues and cause tumors, carbon nanotubes might have the same potential. These and similar immediate concerns about the health consequences of various nanomaterials that are being developed and released into the marketplace are real and require further study. What is encouraging is that unlike the experience with asbestos, whose dangerous propensities were well-known before the public was properly informed, modern standards of scientific integrity, and consumer wariness, are revealing dangers sooner rather than later. This is encouraging unless the scale tips too far to the other side, and unfounded panic supplants safe innovation and responsible science. This is most likely the case with public concern, and hand-wringing by some notable public figures, regarding the so-called 'grey-goo' scenario and its potential to destroy not just humanity, but the world.

Proposed first by Eric Drexler, in his book *Engines of Creation*, it has been repeated by Bill Joy and other doomsayers as a potential (or likely, in the case of Bill Joy) consequence of converging technologies. The scenarios posit that smarter, smaller, self-replicating machines will either become uncontrollable by themselves, and self-replicate using every available piece of matter on earth (until it is a mass of grey goo), or be manufactured to destroy everything by some mad scientist. This sort of nightmare scenario is not new to technological prophesy, as each new technology has at some point been heralded by both prophets and public as the end not just of an era, but of life as we know it. So far it hasn't come true. Even nuclear technology, which has not just the theoretical potential, but actual capability of wiping out the biosphere of our planet, has somehow been contained by either luck or, more likely, common sense and fundamental ethics. Simply put, just because a technology has the capability to be used for evil does not mean the technology should not be developed, nor that it must necessarily or inevitably be used for evil. In the meantime, since its inception, nuclear technologies have been put to significant beneficial use, and prospects remain promising for their future given the threat of global warming from excessive greenhouse gas production.

Nonetheless, in precautionary Europe, and in a smattering of other places tending toward early and expensive bureaucratic consideration of ethical consequences and risks, numerous studies and rounds of hearings have been conducted to try to rein in the grey-goo scenario, no matter how historically unlikely it may actually be. One positive development of media attention and public fear has been its early dismissal as a distraction (UNESCO 2006). Meanwhile, a recent literature study has revealed:

… that as of 2008, seventeen of twenty-four OECD countries surveyed (71%) had developed dedicated strategies for nanotechnology at either the national government and/or agency level. The US, EU, and Australia all have named nanotechnology strategies; the UK also has a dedicated, though unnamed strategy.[10]

Most of the studies being conducted, and regulatory frameworks being enacted, take realistic views of the potential for cataclysmic consequences of runaway nanotechnology. But to what extent will any regulation or other public policy initiative be able to ease the transition posed by such a disruptive technology? If the form of the paradigm shift we might expect from nanowares is correct, then which regulatory or governmental approaches can even begin to anticipate and prepare the public for such changes? Moreover, should they?

I believe that the most relevant and immediately necessary shift in policy that can help prepare the way for a completely decentralized mode of production of material goods would be alteration of our current institutions surrounding intellectual property (IP). It is the disruption in economics, ownership, innovation, and authorship, and our relations with our artifacts, that will be all turned upside down by the best-case scenario of nanotechnology. Yet it is these issues for which we are least prepared institutionally to adapt. Let's consider some of the IP consequences posed by nanotechnology, and explore the hypothesis that this will be the first, biggest hurdle to adopting the technology and encouraging its development to full potential.

Intellectual Property: Unique Concerns of Nano

In the modern era, nothing has both hastened and complicated the landscape of innovation the way that the emergence of IP has. Developed at first by sovereigns (monarchs) as a tool to recruit entrepreneurial activity, or inventive persons, into their employ, 'Letters Patent' and later copyrights were exclusive monopolies protecting various goods and services and their authors or purveyors for a period of time. Letters Patent were used by the English Crown to entice pirates to become 'privateers' (a fancy name for legitimized piracy), by giving them a monopoly over some of the spoils of their piracy for a given time. Sir Francis Drake was employed this way to help undermine the growing Spanish dominion over the Caribbean and New World.[11] Letters Patent evolved slowly into modern patents. At first, they were employed sporadically and less than predictably by monarchs, and later they became part of entrenched and more predictable state institutions. Their modern forms are familiar: patent, copyright, and trademark. The original form of IP protection was simple: it was called keeping a secret. Secret keeping is still used by some innovators where possible. It is cheap, and in some cases quite effective. CocaCola® is a prime example. This recipe has been kept successfully secret for almost a hundred years. It is a valuable piece of IP.[12] But the progress of the useful arts and sciences may be stifled by secret keeping, and secrets may ensure that potentially useful information never enters the public domain except by independent discovery or invention. This is one reason why IP laws were created: to encourage innovation, and ensure that the fruits of invention move into the public domain ... eventually.

The monopolies embodied in patent and copyright laws expire after a specific (though increasingly lengthy) period of time, and the art or invention that was once monopolized becomes common property. Once knowledge moves into the public domain, it can be freely exploited by anyone. For a couple hundred years, the distinctions between types of objects, and thus the sort of IP protection afforded, were clear. Copyrights were for written works, then eventually paintings, photographs, and films. Patents were for inventions. The distinction between inventor and author seemed clear enough. Authors created writings, inventors created tangible objects that did things, or helped us to do things. But recently, this distinction has begun to dissolve.

It was actually software, or what we now broadly call information, communication technology (ICT), that began to undermine the traditional categories of IP law. When software for digital computers first began to be exploited for profits (free software had been the norm for some time, or trade secrets), patents were the first means of protection that programmers sought. This was partly because copyright law used to prohibit granting protection to any form of an 'expression' that could not be directly perceived by humans. The 'direct perception' requirement meant that vinyl LPs (records that spun on turntables, your parents might own a few) and audio tapes could not be copyrighted. The law was changed in the early 1970s to eliminate this requirement, and then software became copyrightable. Suddenly, however, IP met a metaphysical crisis. The categories of patent and copyright had previously been mutually exclusive, meaning one could not patent something that was 'expressive' and one could not copyright something that was 'useful'. Either the nature of these categories was suspect, or software was a 'hybrid' object of some type.

I have argued in my book *The Ontology of Cyberspace*,[13] and elsewhere, that software revealed that the original distinction between copyrightable and patentable objects was arbitrary. The realm of objects covered by IP includes all new 'man-made objects, intentionally produced', each of which is an expression of an idea, and each of which falls somewhere on a spectrum of uses ranging from primarily aesthetic to primarily utilitarian. Even while software revealed this error, which I will expand on and continue to defend later, nanotechnology will finally reveal that our notions of authorship, intention, and object need to be revised to properly deal with the possibilities and prospects of converging technologies.

Nanowares, both in their present and emerging forms of distributed manufacturing, which I will discuss later, and in their future application as a form of molecular manufacturing, involve the creation and distribution not of the final object themselves, but of the 'type'. The type/token distinction in logic, which is mirrored in the 'idea/expression' distinction in IP law, correctly notes the divergence of abstract entities (like the idea of a chair, or the number three) from instances in the world of each. IP protection can only extend to the tokens and not the types. Thus, no one could patent the idea of a chair, but only if it is instantiated in specific forms of chairs (if they are new, non-obvious, and useful). Once one receives a patent for a new, non-obvious, and useful object, the nature of the right extended is unusual. It is not possessory like that given to property-holders. You cannot lay claim to any of the tokens out there of the patented object. Instead, you can prevent others from creating and selling instances of the type protected, unless they pay some royalty. It is an exclusionary right. The same holds true for copyright. The copyright I automatically have on the words on this page prevents others from copying or reproducing them without paying me some royalty (unless, like this book, an open source license enables free access). All of which raises interesting questions of how any of these rights will be applied to objects that will essentially be distributed as types rather than tokens. What will count as authorship of nanotechnology-based objects? How will authors (inventors) of these objects be rewarded? How can the promise of these technologies be realized despite the difficulties of applying traditional IP to their products? All of these vital issues are already confronting emerging precursors of true nanotechnology. We will see in later chapters that developing new models for protecting IP, and applying them to distributed manufacturing in its present and eventual forms, will both serve the needs of economic justice and ensure greater, more democratic means of innovation. Moreover, creativity will continue to be rewarded with justly earned profits.

Ethical, Policy, and Social Implications of Future Nanotech

Disruptive technologies, as I have sketched above, are nothing new. A new and potentially useful trend, however, is approaching the ethical, legal, and social implications of disruptive technologies methodically. Although we can never accurately predict the full impact of any new development, whether it's an artifact, political system, or new mode of behavior expressed through a technology, we can attempt to address new conditions as they arise. Nanotechnology is giving us that opportunity, even as software and the growth of the internet have done so recently.

Major ethical, political, and social concerns raised by nanotech still center largely around two main themes recognized by others who are researching this field and its implications: risks and justice. I am particularly interested in justice. Specifically, the promise of nanotech to achieve technologically what no political system ever could: the end of scarcity. In theory, eventually pretty much anything we need could be manufactured locally at the molecular level, saving tremendous amounts of energy, relieving us of the environmental and economic impact of transportation, and providing everyone with not only the bare necessities, but also what we now consider to be luxury items. Standards of living could rise exponentially in the poorest areas, medicines could be manufactured where and when they are needed with devices that could be accessible to anyone. The dream of molecular manufacturing includes the ability of 'nanofactories' being able to reproduce themselves. The only input required would be some feed source, likely carbon, which can be manipulated into countless forms for nearly any conceivable function. Other feed sources include the molecules in the air around us, and in our waste products, all of which can conceivably be reconfigured, reassembled, and put to use. The technology alone is revolutionary. But when you begin to think of the economic and social implications of such a world, the term 'revolution' is more or less literal.

Our current economic system is built on conceptions of scarcity, needs, labor, and capital that have fed the specialization of labor, and current manufacturing and distribution paradigms for centuries. All of which would be undone if the promise of nanowares is ever fully achieved. Money would mean almost nothing. Surplus would mean nothing. Capital would be unnecessary. Ideas would become the only thing standing between desires and goods. Reconfiguring our economic system to deal with this kind of revolution is the major challenge we face. All of it hinges upon rethinking the relations among innovators, consumers, ideas, and products.

Technologies have altered our conceptions of class before, and proved to be disruptive to both societies and their economies. The industrial age and the computer age are two major examples. In each case, large numbers of people saw old ways of life replaced by new ones. Along the way, some people suffered. Some people never adapted. In each case, the control of these shifts was in the hands largely of those with the capital to invest in new technologies, and political influence to encourage the adoption of those technologies. The profits realized accrued to everyone to some extent. But classes still existed, and in some ways became more distinct. Overall wealth has increased, but we can quibble about the trajectories of individual choice and opportunity. The 'middle class' has grown over time, but large gaps in standards of living remain in the industrialized and now computerized world, and both between the 'developing' world and the 'developed' world, as well as within each.

The promise of nanotechnology, taken to its logical extreme, clearly will upset the established order. Scarcity is part of the engine that drives profits, and desires and needs unmet create markets for those who wish and are able to profit by meeting them. There is considerable risk that those who stand to lose their treasured place in society, and their economic advantage over others, will somehow attempt to either prevent the full promise of nanotechnology from being achieved, or delay it to their advantage. While most of our focus on ethics, society, and nanotechnology has been on potential harms and risks, the greatest danger is that these fears will be manipulated in the public debate to centralize control over nanotechnology's applications, to prevent the full democratic and economically liberating potential of molecular manufacturing, and to ensure that the *status quo* is not disturbed. But it is the nature of the technology, as we have seen with ICT, that it cannot finally be contained. Advances will be achieved with or without regulation, and the only remaining issue will be whether public policy can be guided fairly to help achieve it, or whether it will be used as a means to try to criminalize those who are attempting to deliver its full potential.

This is, of course, the same thing that is happening in ICT. Peer-to-peer (P2P) technologies are a boon to distribution of media, but they got out of control. The media producers, or at least the large, consolidated ones, saw their tight control over the distribution of their copyrighted works slip away as P2P programs allowed the rapid sharing of large files as 'torrents' over the internet. What might have been embraced as an efficient, convenient, and even potentially profitable means of bringing media to more people (perhaps at a reasonable cost) has become the focus of efforts to criminalize it. This likely mirrors what will happen with futuristic nanotechnology. The question is, can public policy and those who want to distribute the products of their creativity figure out a way to embrace the technology rather than attempt to stifle it?

Some people are already beginning to create tools that are intermediate steps between now and the nano-future. These tools are first steps. They include machines that can fabricate locally pretty much any form one can conceive of. There are still tremendous technical limitations, and decades worth of research and development are necessary before true molecular manufacturing can be accomplished, but these tools are beginning to raise the questions posed above.

The Nano-now: What's Currently Happening in Micro-manufacturing and Nano

We won't have to wait until the distant future to discover the complications that arise when innovators decide to try to profit from their creations using distributed manufacturing. There are already nascent forms of nanotechnology, what we might call 'microfab' for now, that are already in various stages of development. These developing technologies are on the nanowares spectrum, and pose opportunities now to explore the issues raised above about authorship, ownership, and innovation.

Among the examples of current and developing microfab technologies is the 'Fab Lab' effort. Developed out of Neil Gershenfeld's course at MIT entitled 'How to Make Anything (Almost)', the idea behind Fab Labs is to create the minimal functional toolset for fabricating just about anything, assuming one can get hold of the raw materials. From the Fab Lab website:

> Fab labs share core capabilities, so that people and projects can be shared across them. This currently includes:
> – A computer-controlled lasercutter, for press-fit assembly of
>   3D structures from 2D parts
> – A larger (4′×8′) numerically-controlled milling machine, for making
>   furniture- (and house-) sized parts
> – A signcutter, to produce printing masks, flexible circuits, and antennas
> – A precision (micron resolution) milling machine to make
>   three-dimensional molds and surface-mount circuit boards
> – Programming tools for low-cost high-speed embedded processors
> These work with components and materials optimized for use in the field, and are controlled with custom software for integrated design, manufacturing, and project management. This inventory is continuously evolving, towards the goal of a fab lab being able to make a fab lab.[14]

The principles employed, and the goals of the program, are the same as those embraced by those who are developing molecular manufacturing. Gershenfeld himself has embraced these goals, but he and his team have, in the meantime, created a large-scale version of the concept. While Fab Labs require users to have some skill in using the tools, the idea has liberated creativity in previously unlikely areas. Setting up a Fab Lab costs about 60,000 USD, and runs on open source software. Fab Labs now exist on nearly every continent, and in ninety locations worldwide. These Fab Labs are an exciting possible front in the war that will inevitably envelope nanowares as it did ITC: the battle between corporate/state control, and grassroots shaping of a 'commons'.

There are other, smaller efforts underway. Desktop manufacturing is the ultimate goal, and so some are creating simple 3D 'printers' that can craft component parts out of various plastics or other similarly moldable materials.

Among these efforts is the Fab@Home project. This is an open source project (meaning that the IP is not controlled by any one person, and cannot be), which aims to develop a simple yet robust 3D printer to fabricate models from computer aided design templates. The Fab@Home website contains a clarion call for the type of revolution I have discussed above:

> Ubiquitous automated manufacturing can thus open the door to a new class of independent designers, a marketplace of printable blueprints, and a new economy of custom products. Just like the Internet and MP3's have freed musical talent from control of big labels, so can widespread RP (Rapid Prototyping) divorce technological innovation from the control of big corporations.[15]

Another promising effort is called 'RepRap' project. This is an attempt to build a truly self-replicating machine that can also rapidly prototype or fabricate any other type of part. The first iteration was called 'Darwin 1.0', and the latest version is being called 'Mendel'. So far, RepRap can manufacture 60 percent of its own parts. The stated goals of RepRap echo those of Fab Labs and Fab@Home. As with all of these efforts, there is a utopian goal of being able to democratize manufacturing and thus liberate intellectual capital in all corners of the world:

> what the RepRap team are [sic] doing is to develop and to give away the designs for a much cheaper machine with the novel capability of being able to self-copy (material costs are about €500). That way it's accessible to small communities in the developing world as well as individuals in the developed world. Following the principles of the Free Software Movement we are distributing the RepRap machine at no cost to everyone under the GNU General Public Licence. So, if you have a RepRap machine, you can use it to make another and give that one to a friend ...[16]

These are all lofty goals, and inspired by utopianism of the best kind. The promise of the technology is clearly the elimination, eventually, of scarcity, and the fulfillment of human needs without the pitfalls of the present economic system (again, eventually). As *The Guardian* reported about the RepRap: 'it has been called the invention that will bring down global capitalism, start a second industrial revolution and save the environment – and it might just put Santa out of a job too'.[17] The same article quoted the founder of Project Gutenberg (which posts public domain content on the internet for free download by anyone), Michael Hart, who properly notes:

> 'In 30 years replicators are going to be able to make things out of all sorts of stuff,' he said. 'Somewhere along this line the intellectual property people are going to come in and say "No we don't want you all printing out Ferraris and we don't want you printing out pizzas".'[18]

What remains missing is the institutional blueprint – the public policies that would need to be embraced, to make this dream become a reality, and to ensure that it leads to a virtuous circle of profit for all. While MP3s have created new opportunities for some artists, there is no doubt that others are 'losing' profits they had expected to receive from their works. Large organizations representing artists, like the RIAA, have fought to regulate the rapid spread of illicit copies of recorded works through P2P networks. Imagine the fight that will erupt if someone posts the complete design specifications of an iPod, and people begin manufacturing them at home on their desktops. This is the inflection point we stand upon: the balance between the great potential, liberating promise of the technology, and the threat this poses to established ways of doing business.

An Outline for the Investigation

Preparing for the future of nanotech requires revisiting some first principles, and then delving into how we might alter currently accepted forms of behavior to meet emerging needs. This approach combines both theory and practice, and I have done this before with both ICT and genes.[19] The first principles involved are those that we use to relate people to both ideas and objects. They underlie our beliefs that, for instance, x is P's idea, and thus P has some claim or right to use the idea. We should look carefully then at the relation of authors to artworks, inventors to artifacts and inventive processes, and the nature of all these, as well as of *ideas, abstract entities, natural laws,* and related objects and concepts. Sorting out how we ought to deal with property or profit in nanotechnology requires first coming to grips with the pre-legal relations that exist among all of the objects and actors involved, then deciding which laws fit the ontology we discover best, and suit our needs most fully.

To do this, we will first look in depth at the technology, from its theoretical inception, to its current forms, including nascent microfab attempts at the grassroots to realize futuristic nanotech. We will then look at the current state of regulation, institutions, and laws, and consider their effectiveness for dealing with the sort of disruption envisioned. Finally, we will look carefully at the nature of property relations, IP, ideas, and people. It is here that we will pave the path for a new way of encouraging innovation. The technological revolution we can foresee contains within it a revolutionary new mode of looking at the nature of IP, although perhaps it isn't so new. Perhaps once again, limitations in the law are being revealed by a new technology, but those limitations were always there.

Nanotech gives us an opportunity to reconsider old concepts and explore new forms of relating innovators, authors, and their creations in ways that both encourage innovation and promise mutual benefit without governmentally supported monopolies. Ultimately we will see that nanowares involve the convergence not just of every other technology, but also of world views. In them lies the germ of an idea that political systems have failed to fulfill: the end of scarcity. It also contains the potential to liberate an instinct that has been necessary for only a limited class of people since the beginning of the industrial era: the creative instinct. When labor became specialized, only a few needed to be innovators, and capital went to those who could raise it on the strength of their good ideas ... sometimes. Many failed. Good ideas alone don't always succeed, as capital has remained relatively scarce for seeing an invention through to success in the marketplace. Now, with the promise of nanowares, and their present iterations in microfab, we might be able to revive the creative, artisan instinct we lost when a broad skill-set became unnecessary. If we can recalibrate, or replace our present institutions, chuck IP law as we know it, and devise a new paradigm for innovation and profit, we just might succeed.

CHAPTER TWO

Nano-futures

Even while we wrestle with the potential ethical, social, and economic consequences of nanotechnology, scientists have far to go to resolve the technical challenges of designing and building things at the nanoscale. This dream, which some call nanomanufacturing or nanofabrication, is still decades away and hindered by the physics of the very, very small. Simply put, and as most who have studied basic physics understand, at the nanoscale, materials are subject to effects that larger things are not subject to. While nanoscale constructions are not fully susceptible to weird and metaphysically challenging *quantum* effects, existing in several states at once for instance, they are nonetheless prone to odd forces, predictable but difficult to manage from an engineering standpoint, in part due to quantum-physical effects. Materials and eventually assemblies that are produced at these scales will be thoroughly unique, and the special engineering challenges posed by designing and building things at the nanoscale are also part of the value-statement for making them. What makes them difficult also makes them interesting and theoretically quite useful.

Technologies do not always develop as we initially dream they will. In the 1950s, visionaries had predicted that the nuclear age would deliver power too cheap to meter, as well as everything from nuclear-powered cars to airplanes. The reality proved far removed from the vision, at least so far. Part of this distance is because of natural limits to the technology itself (an airplane that was powered by traditional fission would require lots of lead shielding, which would weigh down the craft, making the result far from efficient), and the other part may simply be that we must remain patient. The world's first commercial fusion plant is apparently on its way to being constructed, and if it is successful, it will indeed yield safe, plentiful power.[1] Balancing expectations while attempting to promote a new science is often a difficult task, and sometimes over-promoting potential breakthroughs, or over-promising on eventual deliverables, leads to public disappointment, disillusionment, or at worst, discrediting. Nanotechnology has not been immune from overselling, and there is yet a large gap between the promise envisioned by its greatest promoters, and even ardent researchers, and the current state of the art.

We can consider the eventual promise, and even the most outlandish science-fiction scenarios, while still approaching current ethical, social, and legal issues raised by those scenarios. But to do so most fruitfully, we should be clear about the gap between promise and current reality, and also note in which ways our societal responses and planning might work to best marshal the technology from the present to as many likely futures as possible. Managing the development of new technologies is no longer merely a technical or bureaucratic challenge. While governmental grants and regulatory schemes account for a fair amount of impetus and guidance in emerging technologies such as nanotech, both the basic science and commercialization of 'converging' technologies promise to be free-wheeling, entrepreneurial, and more prone to dabblers and 'garage' scientists than previous technologies. In order to help ensure that nanotechnology achieves its full potential, and does not lose favor with funding agencies and venture funds, a careful balance between promise and true potential must be maintained. But what are its true potentials, and how far from the various promises and utopian visions does the state of the art fall? Let's first look at the utopian visions, and then examine the technological hurdles faced by nanotech before finally looking at the state of the art as we slouch toward a nanotech future.

Utopian Visionaries and Dystopian Doomsayers

Perhaps more than anyone, Eric Drexler has popularized the utopian possibilities implicit in the full development of nanotechnology. He wrote his doctoral thesis at MIT, entitled *Molecular Nanotechnology*, which was published in 1992 by Wiley and Sons as *Nanosystems: Molecular Machinery, Manufacturing, and Computation*.[2] This thesis was a continuation of the ideas he presented six years prior in his popular and controversial book *Engines of Creation: The Coming Era of Nanotechnology*, published by Doubleday in 1986. It was *Engines of Creation* that helped stoke public enthusiasm as well as eventual skepticism and fear regarding the potential of nanotechnology. It also generated a scientific backlash and a long-standing debate among researchers working toward developing true molecular nanotechnology (MNT) regarding its potentials, limits, and feasibility. Drexler's vision was inspired by Feynman's famous lecture, 'There's Plenty of Room at the Bottom' and he considers in both his popular and academic works the methods by which useful materials, products, and even machines could be built essentially molecule by molecule. The literal interpretation of this future looks quite a bit like 'Star Trek' and has captured the public's perception of the sci-fi possibilities of nanotechnology.

In this utopian vision of nanotech's promise, any object could be designed and built at whim, from abundant carbon or other resources, alleviating scarcity, unleashing creative abilities, and essentially enabling a world without want. The ecological consequences of such a vision have also recently captured the imagination of those who promote molecular nano-assemblers. If anything can be made on the spot, then the oil-dependent infrastructure necessary now for moving things around the globe largely disappears. The Star Trek vision of 'replicators' that make things for us (and presumably recycle them when we're done with them) is actively being pursued by some researchers, as well as dreamed of by visionaries. Ray Kurzweil has adopted this vision, and included his own techno-utopian spin in which humanity itself merges with these magical machines, and all limits to evolution, and shortcomings of the present human form, are overcome. He adopts the term for this future as 'The Singularity' and promises us that it is near.

Neal Stephenson's novel *The Diamond Age* envisions a similar nanotech utopia, but couches it in a warning about the effects of monopoly and class divisions. Imagining a new Victorian era, dominated by nanotechnology, but limited by central, state-corporate control of 'feeds' necessary to supply the molecular components, Stephenson posits the next stage of the technological and social revolution in the form of 'seed' technology, which eliminates the need for central, controlled feeds, and which could liberate the technology for the masses.[3] Dominating this cautionary tale is the technological vision, populated by remarkable computers, diamond-skinned dirigibles that float and fly by adjustable internal air pressure, and other marvelous technologies that help alleviate scarcity, meet human needs, and satisfy any and all desires. These technological utopias are seductive, and move many in the direction of pursuing the development of true molecular nano-assembly. Science fiction has a tendency, after all, of becoming science fact over a large enough span of time. But for every utopian vision, there is a dystopian response. The dystopian response to Drexler and Kurweil is Sun Microsystems co-founder, Bill Joy.

One technological necessity of some utopian scenarios involves self-replication of nano-assemblies. One radical form of self-replication would be of self-replicating self-replicators. Due to the scale of nano-assemblers, massive numbers of identical nano-robots performing the same task would be necessary for constructing ordinary-sized goods quickly. The value of self-replicating robots was noted by Drexler in *Engines of Creation*, as were its potential dangers. Without some limiting variable, either inherent to a self-replicating robot or within the environment, Drexler noted that things could get out of control. In other words, robots capable of replicating themselves using environmentally available materials could turn the whole world into 'grey goo', according to Drexler. The grey goo scenario has become common fodder for apocalyptic sci-fi as well as in ethical discussions regarding the development of nanotechnology and its sister science, synthetic biology (which we'll come to in a bit). It is technically conceivable, based as it is on a goal of MNT (the development of von Neumann machines – self-replicating robots), but is it practically worrying? Should we be concerned?

Bill Joy was worried enough in 2000 to publish his dark article 'Why the Future Doesn't Need Us' in which he predicted the likely extinction of the human species due to our technology. Of course, a number of scientists have pondered publicly the potentially apocalyptic possibilities of numerous technologies, and fear about the demise of earth at our own hands has become somewhat commonplace, including the very real public angst associated with the scientific likelihood of catastrophic, anthropogenic global warming. Joy's article essentially warns that the perfection of the three most promising technologies of the twentieth and twenty-first centuries, genetic engineering, robotics, and nanotechnology, would mean the development of machines that could well inherit the earth, with no need for us. The *Terminator* scenario of robots destroying their creators, fears about man's inventive hubris and runaway technologies in general, dates back at least to *Frankenstein* and even earlier, to *Faust*. At the core of these concerns is an element of reality. Our technologies sometimes do pose hazards, and the unintended consequences of the most benign inventions (see, e.g. opiates and other addictive medicines, or even fossil fuels) sometimes do sneak up from behind and bite us. Bill Joy's angst might be overstated, and his fears may be quite premature, but they are worth keeping in mind as our technologies become more sophisticated and more difficult to control.

Fortunately (or unfortunately) for now, a number of technical hurdles stand in the way of the most utopian visions of nanotech, and overcoming these impediments affords us some time to do a bit of philosophical self-reflection and moral perfection before the potential of a grey goo scenario becomes imminent. Molecules behave differently than the scale of objects ordinarily used in day-to-day engineering, and just creating objects at the microscale is difficult enough. At the nanoscale, forces are unfamiliar and will require new modes of design and manipulation to overcome, pervade, and control. Although the sci-fi future of nanotech is likely a long way off, it is not impossible. In fact, technical hurdles tend to be overcome, in time, and Star Trek-like replicators may someday come to be. In the meantime, however, there are some transitional technical and social steps that are being undertaken to realize some of the utopian vision. We might also posit that these dystopias are similarly possible, if we look far enough into the future. Scientists must first perfect the science before the engineering of nanoscale machines, much less self-replicating nano-robots, could become a real technical possibility.

Special Problems of Manufacturing at the Nanoscale

Richard Smalley was among three physicists who collaborated on one of the great practical breakthroughs in nanotechnology, or more specifically, in nanomaterials research. Robert Curl, Harry Kroto, and Smalley comprised the research team that was awarded the Nobel Prize in Chemistry in 1996 for their discovery of buckminsterfullerene, popularly known as 'buckyballs' for their geodesic dome shape. This newly discovered though ancient form of carbon has numerous envisaged applications in materials, drug delivery, power storage, and other uses. Buckyballs and carbon nanotubes (a stretched out version of the buckyball) have remarkable strength and flexibility, and interact with other atoms in ways that may make them particularly useful for a variety of applications. Smalley was also an early and vocal critic of Drexler, citing what he called the 'fat fingers' and 'sticky fingers' problems as obstacles to molecular assemblers. His public debate with Drexler, published in Chemical and Engineering News, also cites what he considered to be the dangers of over-promotion of the potential of the technology, and apocalyptic doomsayers like Joy spread similarly unlikely prospects for such a young and undeveloped technology.[4] The fat and sticky fingers problems, however, and similar technical and physical hurdles associated with materials at the molecular scale, stand as real obstacles to nano-assemblers, and will require a significant scientific and engineering leap-forward before the utopian or nightmare scenarios discussed popularly are ever to be fully realized.

The essence of the practical problems associated with nano-assembly is that molecules are extremely small, and any assembly built to interact with individual molecules will face the problem of 'fat' fingers. Try, for instance, using your fingers to manipulate the screws that hold your eyeglasses together. Smalley described this and the problem of sticky fingers in an article in a special issue of *Scientific American* devoted to nanotechnology and published in 2001. His article was entitled 'Of Chemistry, Love and Nanobots'.[5] Reacting to the article, Drexler and others describe theoretical means of overcoming the problem of manipulation by citing the successful manipulation of single molecules within a vacuum, and using a substrate to bind the substrate (upon which the single molecule is bound) as, for instance, had already been successfully accomplished in a scanning tunneling microscope (STM) at that point.[6] They also point to an alternate approach in which larger components are used and assembled and used to make assemblers, which in turn would manipulate larger parts. Presaging the current popularity of 'synthetic biology', they cite the assembly of proteins by mRNA in cellular ribosomes, and note the molecular precision of this natural mode of nano-assembly.

'Sticky fingers', according to Smalley, regards the stickiness of molecules to the manipulator, whatever it would be. In order to manipulate molecules into position, an assembler would need to bind with the component molecules, and somehow precisely place them in whatever is being assembled. Yet the stickiness necessary to pick up and move the molecule into position would have to be overcome when the molecule or molecular component has been placed. This is a 'fundamental' problem, according to Smalley. Drexler *et al.* respond that, once again, molecules have been precisely placed already in an STM, with the application and discontinuing of current to achieve the placement. Moreover, they once again cite the example of the ribosome as a precise molecular manufacturer, and raise the possible uses of a scanning probe microscope for precise placement of molecules in an assembly. These debates will be resolved not in the pages of journals, but by technologies that either eventually address the real, physical hurdles faced by MNT, somehow work around them, or never materialize.

Other hurdles to overcome have been pointed out by Richard Jones, physics professor at Sheffield University, in the United Kingdom, in his 'Six Challenges for Molecular Nanotechnology',[7] which he poses not as roadblocks, but as scientific challenges that *can* be overcome, as scientific challenges often are. These challenges include maintaining the stability of new molecular configurations (they like to bend and morph in ways that are often unpredictable, based upon nature's tendency to seek equilibrium). A second challenge is to thwart the effects of environmental heat and Brownian motion. As we learned in high school physics, very small things are influenced by even smaller things, like vibration caused by heat, or being struck by smaller particles in the environment. Heat is a factor in the third problem that Jones identifies, namely problems associated with friction in nanosystems. Because of the (relatively) large surface areas involved, and the degree to which additional heat and dissipation of energy through friction may affect such small systems, better models are needed than those currently used to design and engineer systems that operate predictably. Jones also notes that a molecular-scale motor poses design challenges not posed by larger-scale motors, specifically issues relating again to surface stability and the tendency to adhere (rather than move). A fifth problem Jones notes is the creation and maintenance of a 'molecular mill', in which molecules are selectively taken in from a feed or the environment, and rearranged to become part of a larger product, all in a nearly perfect vacuum. Finally, Jones notes a larger engineering problem: that of transitioning into a true MNT engineering scheme. Drexler and others point to biological processes for examples of functioning nanotech-like, natural mechanisms. Developing such systems is now the project of those engaged in 'systems biology' or 'synthetic biology',[8] and is indeed reaping some early rewards. But moving from that to a 'diamond-based' or other true molecular nanotech future currently lacks an engineering middle-ground, or transitional phase. The chemistries are apparently fundamentally different.

What do these challenges imply? Is futuristic nanotechnology, of the Star Trek replicator sort, fundamentally unachievable, or will it just take more than a few decades to solve these engineering problems? I assume that it is the latter and not the former. Unless physics dictates that any of these challenges cannot be resolved or worked around, we should assume that somehow they will be overcome ... eventually. None of these present fundamental physical barriers. New technical achievements are already being made in nanotechnology, and now in its sister (perhaps precursor) science, 'synthetic biology'. So let's look first at what is being achieved and put the technical hurdles that nanotech still faces in perspective, before we begin to consider some of the social, ethical, and legal challenges both posed by and that may be imposed upon nanowares and their future development.

Nanotechnology Achievements: The Nano-now

Spurred in part by increased public funding for the basic research, nanotechnology now boasts numerous scientific breakthroughs, and an increasing number of commercially viable applications. These achievements range across nearly every conceivable domain, from environment to energy, health, leisure, food, and infrastructure. While we likely have a long wait before 'true' MNT robots constructing anything we dream of at whim, nanotech scientists have made significant scientific advances, and hundreds of consumer products already boast some or all of their capabilities as the result of nanotechnology. Most of these breakthroughs have occurred not in MNT, *per se*. There are no nanobots autonomously scurrying around making things for us, cleaning our bloodstreams of cholesterol, nor any of the infinite number of sci-fi uses to which we hope, eventually, to put futuristic nanotechnology. But significant advances have been made in nano-materials. Scientists like Kroto, Curl, and Smalley, whose discovery of buckyballs ushered in much of the current spate of nano-materials research, opened the door for further investigation of the interaction of nano-particles of various kinds with other types of materials, so that now we have the option of using sunscreen, paints, or even consuming foods that owe their improved properties to the use of nanowares of some sort.

Scientific advances achieved in 2009 with funding by the US National Nanotechnology Initiative, for instance, list studies involving the use of nanoscale powders and materials involving aluminum and other metals, polymers, carbon and protein nanotubes, and numerous others. Nano-materials research carries none of the technological impediments or challenges discussed above because various macro-scale processes can be used to create nanoscale materials. They need not be constructed by little robots. But nano-materials have a number of features that make their industrial uses attractive, and that are proving promising for helping to solve important engineering problems. In energy, for instance, nanoscale materials are improving the storage capacity, lifetimes, and efficiency of batteries. Nano-materials are improving the capabilities of air conditioners, heaters, filters, and purifiers. They are being used in detergents, sunscreens, deodorants, paints, home electronics, and to help ensure the safety of foods, as well as to cut down on the spread of bacteria through surface applications in kitchens and around homes. These materials are both interesting and useful because of their surfaces. As with all nanotechnology, it is the interaction of nanoscale objects with each other and their environments that is facilitated by the large surface areas involved, and the special characteristics of materials at that scale have numerous engineering applications.

For carbon-based nano-materials, such as 'fullerenes' (which can be made in sheets, or remain as balls or rods), useful properties include a very high strength to weight ratio, as well as electrical conductivity. For 'non-organic' nano-materials, including most usefully various metals, practical features include high conductivity, or even superconductivity, high mechanical resistance, and chemical inertness. Chemical inertness is important for use in foods and medicines, and these are the current targets of application of many of these new materials. As well, their high surface areas make them useful in or as catalysts, helping to speed reactions by reacting with greater volumes due to their high surface areas. There are over 800 commercial products now listed as involving some material that could properly be called 'nano' in some regard. The Project on Emerging Nanotechnologies, funded in part by the Pew Foundation, and located at the Woodrow Wilson International Center for Scholars, keeps a publicly available list of such commercial products.[9] They also publish occasional reports on social, ethical, and legal issues relating to nanotechnology, in both its future and present forms.

The growing prevalence, and perhaps eventual ubiquity, of nano-materials in consumer products has stoked concerns about health and safety. This concern is rooted in past errors in judgments, or lapses of scientific knowledge, regarding materials that ended up in foods, medicines, and household products before they were found to be dangerous. Witness, for instance, thalidomide, asbestos, DDT, lead, and countless other once commonly used materials thought to be safe, but later found to be quite harmful. Given this history, it is not surprising that the flow of nano-materials from lab to market is causing some concern, and policy-makers are being increasingly asked to provide some oversight. Indeed, oversight and caution are warranted with the introduction of any new technology into commerce. The special natures of nano-materials may make them dangerous in ways in which scientists have not yet predicted, just as they may have uses which are similarly not predicted. Asbestos, for instance, is shaped similarly to nanotubes, and it is the shape of asbestos particles that makes them particularly capable of causing cancer if inhaled. So we should study the potential risks of nanotechnology and test new products as we do with all new, untested products meant for human consumption, or used in medicines, or with potential for environmental exposure. But while nano-materials may have unique characteristics requiring risk assessment and oversight, the ethical issues involved are nothing new. The ethics of risk and the social and moral responsibilities of scientists and engineers making products for human use are well-established, and are not particular to any one science or technology. Even so, we should look carefully at the distinct technical natures of nanotechnology in general, consider in-depth the philosophical implications (for some are unique), and see how these issues ought to inform policy.

To do so accurately, and most usefully, we should distinguish between the nano-now (what is possible and being achieved now) and the nano-future (with all the sci-fi implications of fully developed MNT) and see if there exists some path between the two that might be informed by policy considerations. I have always approached these sorts of issues by looking first to the nature of the technology, and asking whether something unique about that technology requires special policy consideration. In other words, we must first carefully analyze the *artifacts themselves*, and then move upwards toward social, ethical, and legal implications. As I mentioned earlier, most of the risk-related ethical challenges posed by nanotechnology are not unique. They are challenges posed by any risky new science and technology, and they involve the same duties that scientists and engineers owe in the development of any new object meant for public consumption or environmental release. But there is a way in which nanotechnology poses unique challenges. This is because nanowares of the MNT variety are fundamentally transformative technologies. If true MNT is developed, or more optimistically, *when*, then the nature of our relationships to the world of objects will be changed. Our relationships to each other, as innovators and consumers, for instance, will likewise be forever changed. The germ of this change can be seen in the last transformative technology whose full impact is still not fully realized, and which has altered our societies and economies. Information and communication technologies (ICT) have changed our world, and ushered in a host of social, ethical, and legal challenges to every society, many of which are still being grappled with.

I have argued at length that taking an ontological perspective first, by which we first accurately describe the nature of technical artifacts (such as ICT or genes), is the best way to design policies that make logical sense. Thus, for instance, a system that tolerates both the patenting and copyrighting of software is illogical, as well as inefficient.[10] Granting patents to unmodified genes is similarly nonsensical given that they belong to a 'commons by necessity', which is any natural space that cannot be enclosed in any meaningful way.[11] So what are the ontological implications of nanotechnology? How do the objects and processes involved with it make it potentially transformative in ways other technologies have not been, and how might its unique characteristics and an accurate ontological picture shape our policy decisions? Besides the very real but not entirely unique ethical issues posed by risks (which will be discussed more fully later), what is the *nature* of nanowares, and what does this mean for us and our relationship in general with the worlds of artifacts both nano- and macro?

Nano-artifacts: Some New Philosophical Challenges

Although the ethical dilemmas and challenges posed by true nanoscale artifacts (including materials and eventually nano-machines) are generally no different than those posed by any other technical artifact, there are some interesting issues posed by what amounts to 'programmable matter'. Specifically, we have for a long time labored under a preconception about the nature of all artifacts, technical and otherwise, based upon some ancient philosophical distinctions that bear revisiting. These preconceptions have been challenged already, with the emergence of ICT, and nanotechnology now promises to finally destroy them. They are based on Plato's metaphysics, in part, and infuse our laws in some important and confusing ways. Most of us adopt these distinctions as well in our everyday parlance, but following them too literally or inaccurately in the law and public policy has resulted in a strange state of affairs that I noted in *The Ontology of Cyberspace*, and warned this would prove to be a problem for nanotechnology. Now, let's settle it.

Plato conceived of the universe as existing in two planes. The world we experience and the world of 'ideals'. To summarize, the world of ideals is the perfect realm, in which the perfect form of every worldly object (including abstract objects, like virtues) exists. The world of experience is but a shadow of that realm, and philosophers strive to grasp the forms of the ideals. The well-known allegory of the cave, from Plato's *Republic*, illustrates our relationship with the realm of ideals or 'forms', as we inhabit the cave, and only see shadows cast by the most real world (the realm of ideals) into our imperfect world (the cave).[12] The necessity of philosophers, whose grasp of ideals becomes greater than that of ordinary people, makes their position in society necessarily privileged. One can see the appeal of this theory to philosophers, who have, in one way or another, perpetuated the allegory even into modern discussions about mind and matter, consciousness and reality, and discussions concerning abstract versus concrete objects. Today, we still grapple with the distinction between what is real and what is not, but too often, we consider ephemeral things, that after all do exist, as being equivalent to 'spirit', 'soul', or other intangibles. Lack of understanding of both the objects of reality, and the language used to properly describe it, causes real legal, social, and political problems. The Supreme Court of the United States, for instance, recently repeated the unfortunate term 'too abstract' once again to limit types of patent.

The law of intellectual property (IP) is replete with the sloppy thinking that results from inexact understanding of what constitutes an idea, and how ideas differ from other material objects. Consider, for example, a recent Supreme Court case that was meant to clarify the issue of whether certain processes (specifically, business methods) could be patented. In *Bilski v. Kappos*, 561 US ____ (2010), the Supreme Court did little to clarify the question of what counts as a patent-eligible 'process' under the US Patent Act (which allows patents on any new, useful, and non-obvious 'process[es]', 'machin[es]', 'manufactur[es]', and 'composition[s] of matter'), or to clear up the ambiguity implicit in the string of court cases that prohibit patent eligibility in the case of 'laws of nature, physical phenomena, and abstract ideas'. Unraveling the metaphysics implicit in patent law reveals the depth of our Platonic preconceptions, and sorting out this quandary uncovers confusions that obviously continue to befuddle the courts. Setting aside the pragmatic goals of patent law, let's first try to grasp the objects involved, and then see how they may or may not be properly described in the law.

The owner of a patent has the right to exclude anyone else from the production or practice of the invention claimed. So, let's consider the above proscriptions as applied to a tricky example, and then see how the courts have hopelessly confused things. Consider Joseph Priestley, who in 1774 discovered oxygen by heating mercuric oxide. He found that the gas released from heated mercuric oxide was quite combustible. He had isolated pure O_2, gaseous oxygen, from a compound. What are the potentially patentable parts of Priestley's activities? There is the process of separating oxygen from mercuric oxide, and there is the product (pure O_2) that is obtained. Under current US precedent, an argument could be made that Priestley could obtain a patent for both the new, useful, and non-obvious process by which the O_2 was created (or isolated from its previous compound state) and also for the product as created by this process (O_2). Patents have been applied for and obtained for the elements americium and polonium, both of which are radioactive heavy isotopes that are generally produced by man (although natural processes might make them elsewhere). Patents have also been granted for synthetically produced analogues of naturally occurring products like insulin and adrenaline. These patents cover not just the processes of creating them, but also the products. So, sorting all of this out requires a rather challenging interpretation of what counts as a law of nature, a physical phenomenon, and an abstract idea, one that, as we'll see, falls apart under scrutiny. All of which poses a particularly interesting challenge to a future in which the material world becomes programmable at the molecular level.

This is how it works out under current law: O_2 is a product of nature, and thus not patent-eligible in its natural state. The *reason* it is not patentable may be that it is either a 'law of nature' (which it seems it cannot be, since it is a *product* of nature) or perhaps because it is a 'physical phenomenon' (which seems more likely). O_2 is certainly not an 'abstract idea' since it exists in the universe as something we can experience directly, and do so with every breath we take. The process is less problematic, though it is still somewhat tricky. It is certainly conceivable that mercuric oxide, which exists in the environment in mineral form (as montroydite), has been heated naturally in the past, and released pure O_2 in the process. But humans and other creatures have been producing adrenaline and insulin for eons as well, and yet the courts have held that the 'isolation and purification' of such natural products constitute sufficient inventiveness to warrant a patent. So what then do the courts do to distinguish O_2 as a physical phenomenon from O_2 as a product of man? The genesis of a particular O_2, insulin, or adrenaline molecule must be what matters. Thus, two structurally identical molecules are different, according to patent law, if one was produced by some human intention, and the other produced through some naturally occurring process. For patent law, structure *and* origin of the object matter. If Priestley had obtained a patent for O_2 under current precedent, no one else could isolate O_2 in any manner, since the patent would cover both the process and the product. So even were you to discover that O_2 could be separated from water by electrolysis (it collects at the anode of an electric cell in water), as William Nicholson and Anthony Carlisle discovered in 1800, the Priestley patent would foreclose your new production technique, although you could get a patent on the new, useful, and non-obvious process.

According to the current legal ontology of patent, an object that is morphologically identical to another may yet be considered to be different in a legally significant way allowing the patent on one but not the other. All objects thus must have a structural quality, and a genetic quality, and if both are the result of some human intention, and meet the other criteria of patent (new, useful, and non-obvious), then they may be patentable. How this relates logically to the exclusionary qualities 'natural law, physical phenomena, and abstract ideas' is unclear. It is important that we sort this out, because the future of nanotechnology rests upon the production of things at the molecular level, and a complicated web of intentions. Extrapolating from the current state of the law, which calls two identical objects importantly distinct due to their genesis, results in an impossibly complex nanotech future in which each new nanotech component could become patented, and tracking the ownership rights of any useful nanotech-based artifact would become a *pragmatic impossibility*. Besides the practical problems posed by this scenario, we might well ask whether the philosophical foundations are sound.

What counts as an artifact? This is a long-disputed philosophical problem. From a realistic point of view, the world consists of at least two types of objects: artifacts and nature.[13] I have argued that anything in the world which is intentionally created by man is an artifact.[14] But without elaboration, this leaves a fair amount of overlap between artifacts and things we might not necessarily consider to properly be artifacts. Priestley's O_2 would count as an artifact, since he intends to create O_2. Yet O_2 exists in nature, even in isolated and purified form as the product of photosynthesis, by which plants strip the O_2 off of CO_2, synthesize the C for their growth, and excrete the O_2 as a waste product (luckily for us). So is 'synthesized' O_2 an artifact or is it a product of nature, *artifactually* created?

There are millions of natural phenomena which are duplicated by man. Consider fire. Fire occurs in nature, and yet we much prefer fires that we create and control to those created naturally. The 'thing' fire, which is the plasma state of matter undergoing combustion, is therefore in our ordinary experience (with gas stoves, for instance, or fireplaces) a natural product harnessed in some intentionally generated and controlled process. Each instance of a man-made fire is the reproduction of some natural thing (fire) by means of some artifactual process (like lighting a stove, containing it in a particular place, feeding it with fuel, etc.) which does not alter the fact that the thing itself is natural. It is a physical phenomenon, which should be an exception to patent eligibility. Does its genesis in human intention matter? It does, and we can recognize the artifactuality of the genesis without conflating the object with its process. In fact, by conflating products and processes we commit a grievous ontological error, as there is no commensurability of products and processes. Things (continuants) persist and are extended in space, and processes (occurrents) act upon them.[15]

O_2 is O_2, whether it was created due to human intention or not. The *process* of generating O_2 from some intentional act, like heating mercuric oxide, or through electrolysis, is itself an artifact, a man-made intentional thing, which deserves consideration as patent-eligible. Claiming more, as for instance extending the artifactuality to the product as well as the process, is not ontologically warranted, and, as we'll see, becomes a practical problem for IP law, not to mention a nightmare for unraveling ownership issues in a nanowares future. The creation of fire by human means does not create a new type of *fire*. If this were so, then each instance of fire would have to be characterized by its particular means of generation. This would be a 'match fire', that would be a 'flint fire', the other would be a 'lightning fire', etc. The world should not be populated by so many objects. There are fires, and they owe their origins to different causes, but the physical phenomenon is the same. The occurrents through which each continuant is altered differ, but identical continuants ought not to be split into different sorts of objects due to the acting on them of various differing occurrents. This would be a muddied ontology, and an overly complicated world.[16] Yet this is exactly what the law as it currently stands demands of us.

Moreover, IP distinguishes between products and processes, and one must make claims in one's patent application for each separately in order to receive protection. Products and processes are mutually exclusive categories. No product is a process, and vice versa. This much the patent law gets right. We can recognize, in granting a patent for a new, man-made, useful, and non-obvious process for creating an otherwise natural product, the genius of the inventor, and encourage and reward that invention with a patent on that process even as we deny that the *product* is patent-eligible. The primary reason we should deny it as patent-eligible is simply that it is not in any way *new*. We need not delve into the various court-recognized exceptions, such as laws of nature, natural phenomena, nor 'abstract ideas'. Two of these exceptions simply reiterate the requirement of 'newness'. O_2 is not new, it has been around for billions of years, is abundant in our environment, and makes us live. Laws of nature and natural phenomena are, by nature, as old as nature, and pre-date human inventiveness.

All objects are either products (continuants) or processes (occurrents). There are two types of each: man-made and natural, for the purposes of IP law. But this is too simple, as the problem of O_2 makes clear. We must elaborate. There are man-made objects, intentionally produced (all of which might properly be called *expressions* – the manifestation of some intention). There are accidents, or man-made objects with no intentionality behind them (a sneeze, the metal filings left over after constructing something with a lathe). And there are natural objects and processes, whose existence and form occur by virtue of natural laws or processes, or that are those processes. Now let's clarify where the law and our folk understandings of the world have confused matters, and threaten the full realization of nanowares.

Can an Idea be Other than Abstract?

No. Ideas are by nature abstract. There is no such thing as a concrete idea. Ideas can become instantiated in something concrete, in which case they are no longer ideas, but rather expressions. Neither can anything be *too* abstract. The abstract and the concrete are mutually exclusive categories, there is no spectrum. *Pi* is an abstract. A particular circle is concrete, and embodies the ratio *pi*, but no one circle can ever *be pi*. Laws of nature are similarly abstract, although they are instantiated in every natural phenomenon, including both natural products and processes. Ideas are also abstract, although thoughts are not. When we talk about ideas, we are talking about a very basic distinction between types and tokens. An idea is a type. Thus, the idea of justice, the idea of a container, the idea of redness, the idea of steam engines, the idea of relativity, etc. are all what we might also properly call *universals*. These are abstract, in that they do not themselves exist, but rather come to be instantiated in tokens.[17] Some tokens are even thoughts, so the idea of justice becomes instantiated in our own particular conceptions existing in our minds of something just. The idea and the thought are often confused, and this is part of the legacy of Plato. The 'realm' of ideals (or ideas, as sometimes it is called) is no realm at all. Aristotle understood this, and did not literally conceive of a realm in which universals exist more perfectly than things in the real world do. Rather, he conceived of universals only existing in the particulars in which they come to be instantiated. Substance, rather than ideals, was the foundation of all reality to Aristotle. Unfortunately, Aristotle's realism has not taken as firm a hold of popular and legal conceptions of artifacts as has Plato's, and courts still use pleonasms like 'abstract ideas'.

Thoughts are not abstract because they exist as states of affairs pertaining to neurons in brains. The ideas they instantiate are abstract, as they are universals. So, to clarify the confused ontology that courts have created, the real distinctions among the relevant objects are types (ideas, ideals) versus tokens (particulars, instances), continuants (products) versus occurrents (processes), and artifacts (man-made objects, including processes, intentionally produced) versus accidents (man-made objects or processes without intention behind them, or toward them) versus all things natural (objects, products, processes, phenomena, laws, that are part of the fabric of the natural universe, having nothing to do with human intention). An idea cannot be 'too abstract' as some courts claim when rejecting certain patents. Ideas are *necessarily* abstract. Thus, it is redundant to say 'abstract idea'. All natural things are also necessarily *not new*, even if we newly discover them. They might be new *to us*, as nuclear fission was when we harnessed it into a nuclear explosion, or in a nuclear reactor, but the phenomenon of fission has been around for a long time before us, and will continue once we are gone. Similarly for O_2, or naturally occurring genes (as opposed to engineered ones), and the proteins they help create.

So then, from a logical standpoint, as opposed to questions purely about practical necessity, what ought to be eligible for patent, and what ought not? The law already distinguishes between ideas and expressions, and no ideas can get a patent. Rather, the particular instantiation of an idea into an expression may be patented. Which means that the idea, for instance, of using the click of a mouse to select an item for purchase (like Amazon's 'one-click' patent)[18] is not either 'too' or 'sufficiently non' abstract. It is abstract. It may be instantiated in a particular algorithm that accomplishes this, but as it stands, it is an idea, and not an expression. A particular algorithm that accomplishes this, and instantiates the idea, would be properly patentable, but it would leave open the potential for any number of other algorithms to accomplish the same idea. A single idea can often be instantiated in numerous particular forms, as the universals 'redness' or 'justice' can take on a large range of potential forms or instances. So, new, non-obvious, useful expressions of ideas could be patentable, but the ideas themselves never could be, as all patent attorneys will admit. Unfortunately, under the current interpretation of patent law dominant among courts and lawyers around the world, ideas actually *are* being patented. The one-click patent ought to be limited to the particular algorithm used to accomplish the idea of ordering with one mouse click, but it forecloses others from achieving the same idea using different particulars. No patent could issue for a string of nucleotides that occurs in nature, although they now frequently are. The reasoning behind patents on naturally occurring sequences is based upon confusion of product with process, and unnecessary multiplication of entities. The process of discovering the string, or even of replicating it in the lab, might be new, but the string itself – the product – is not new. Neither is O_2 ever new, although the means by which it might be isolated may be.

This view would significantly narrow the scope of any particular patent, but it would at least be consistent with the stated limits of patent law. Every patent lawyer will tell you that no idea can be patented, although actually (and this they won't admit) many now are. Natural laws, phenomena, and abstract entities are increasingly (and inadvertently, perhaps) awarded patents largely because the courts have an inexact understanding of the relevant concepts and terms, and patent lawyers who wish to protect their clients to the maximal extent have used this confusion to expand the law's scope beyond that for which it was originally conceived. If the law were logical, then products and processes could not be conflated, and artifactual processes that create natural (and thus, non-new) things could be patented although the product could not. If courts and patent offices consistently applied the underlying precepts of patent law to actual patents as suggested above, then patent law would not prove to be much of a threat to the development of nanowares. Unfortunately, the current state of affairs in patent law is inconsistent and irrational, and will only hamper the development of nanotech as it has hampered the development of software and gene-based technologies. Fortunately, some scientists understand the vital distinctions outlined above, and Smalley, Kroto, and Curl patented (in US Patent 5227038) only their new *process* for creating 'buckyballs' and not buckyballs themselves. But using the flawed reasoning I have outlined above, by which numerous patent attorneys today would suggest that Priestley could have gotten a patent on O_2 because the process used to make it was *artifactual*, buckyballs might have become patented, and an important basis for much of what is happening in the 'nano-now' would have been foreclosed, or at least become more expensive, to researchers around the world.

When Software and Hardware Merge

The problems that have beset software have impeded genetic research, synthetic biology, and will prove problematic for nanotechnology as well. These problems result from an illogically over-expansive interpretation of patent eligibility, and of conflation of subject matters of IP law in general. A system that was designed when texts and machines appeared to be very different from one another has broken down in light of the fact that, in fact, copyrightable and patentable objects are of the same substance, and serve overlapping purposes. With the rise of ICT, and the development of software, the confused ontology of IP law in general has been revealed. All man-made objects (including processes) intentionally produced are expressions, and the subject matter of both copyright and patent is always some type of expression whose purposes vary, but which actually overlap. Copyright was devised for primarily *aesthetic* expressions, and patent for primarily *utilitarian* ones. But there is no bright-line distinction between the two and software reveals that all expressions ought to be covered by a single scheme, as their objects are of the same type.

This would, of course, require re-interpreting IP law, but only in ways which would finally be consistent with their stated underlying principles. Ideas (universals) would be treated differently than expressions (tokens, particulars), and products would be recognized as ontologically distinct from processes. Each patent would have to reveal the particular form of solving a problem, instantiating the ideas precisely in some functioning manner, and the scope of the patent could not extend beyond that form. Amazon's one-click process would cover Amazon's one-click process (the particular algorithms used to achieve it) and not extend beyond that token to the type (such as to any other one-click process). A patent could never intrude upon the type, nor could it extend into the territory of *the natural* (and thus, not new). While the particular application of a natural law to a new expression might be rewarded, or a new method of producing something not new might be protectable, nothing beyond the man-made part of the process or product could be foreclosed from use by others by way of patent.

Because the courts and patent offices have failed to abide by these principles, innovation has suffered, and authors and inventors must be particularly wary when creating anything new, or even when applying what they once considered to be old (for instance, because it involves some natural product or process), because some patent might well have claimed a monopoly for a particular product or process. Software authors have reacted in many ways, including by devising new methods outside of the traditional IP regimes to develop their software, by avoiding the potential for litigation by making their products free, or by adopting open source licenses. In synthetic biology, a similar course has been taken through the BioBricks Consortium, and other models of collaboration by which the constituent parts of synthetic biology (which will be explained in more detail in subsequent chapters) have been excluded from patentability by agreement, by revealing and sharing those constituent parts among the community of developers, rather than being patented or otherwise protected by law.

If IP law continues to be misapplied to nanowares, it will continue to thwart innovation and complicate growth of this promising new technology. Perhaps recognizing this potential, numerous innovators who are building technologies that promise to bridge the gap between the nano-now and the nano-future are adopting mechanisms that skirt traditional IP, and that keep the basic building blocks of nanotechnology open. Perhaps the most revolutionary element of future nanotech will be the ultimate decentralization of manufacturing, freeing up products to move anywhere on earth, unencumbered by traditional supply chain issues, vastly reducing the amount spent on manufacturing and transportation infrastructures. In the utopian and dystopian visions of nanotech's promise, items can be designed anywhere, and delivered anywhere, assuming the possession of whatever nano-assemblers are eventually used. Assuming also the ubiquity of the materials used to fabricate things in such a future (such as carbon, which is both abundant and capable of being made in numerous useful forms – diamond, nanotubes, graphene and other fullerenes, etc.) the net result of such a future is the end of scarcity. It is this vision which is driving many of the current, open source, intermediate steps that may usher in many of the most evolutionary facets of the nano-future, and which has made amazing headway, even without traditional IP to support its various innovations. In the next chapter, we will look in-depth at some of these efforts, and how traditional IP hampers innovation for the nano-present, as well as why innovators are looking beyond patents and copyrights to promote innovation.

Case Study

The Diamond Age

The scenario:

Neal Stephenson's novel posits a plausible far-future vision of nanowares, in which true MNT is (nearly) fully realized. As with the dystopian vision of his previous novel, *Snow Crash*, the warning innate to *The Diamond Age* concerns not the technology itself, but who controls the technology. The story consists of two intersecting plots centered upon a nanotechnological book (*The Primer*), which may or may not also incorporate true artificial intelligence, but which nonetheless contains the key to altering the nature of the society in which the characters live. That society is built upon the ubiquity of MNT, but only of a certain kind. The infrastructure by which 'matter compilers' construct and deliver products is centrally controlled by the ruling class (some combination of the corporate elite and the state, or states). The technology is thus managed by a central, hierarchical order. While some of the promise of the technology is realized, ensuring that anyone can get some basic necessities for free at public matter compilers, the stratification of society is maintained by control of the 'feeds' that supply matter compilers with basic molecules. Public matter compilers can only deliver basic goods, and control of the feeds ensures that the ruling class continues to profit by delivering both component materials and designs for final products to those who pay for them through private matter compilers. None of this fully prevents illicit uses of the technology, as nanoware wars and terrorism are a constant threat, and nanoware defenses continually maintain the safety of citizens.

In stark opposition to the 'feed' technology that fuels the neo-Victorian society of *The Diamond Age* is the illicit knowledge

---

Case Study: *The Diamond Age* (continued)

of 'seed' technology. The suppressed 'seed' technology would liberate the production of nanowares from the 'feeds' and enable production of basically anything, anywhere. It would be the full realization of the Drexlerian vision of MNT, and its implications for the hierarchical societal and economic structure of the novel are clear. The suppression of 'seed' technology and dependence upon the centrally controlled feeds enable the means of production to continue to be controlled by those who want to profit from artificial scarcity. Maintaining scarcity is essential to maintain the social order as it is. 'Seed' technology threatens revolutionary social upheaval, and the breakdown of hierarchies built upon scarcity and wealth.

Questions presented:

Neal Stephenson's novel presents two plausible directions of efforts at creating nanowares, both technologically and socially. Can we picture the current regulatory, legal, and institutional trends as progenitors of Stephenson's visions of 'feed' versus 'seed' technologies? How might current legal, regulatory, or other institutional structures fit with the two potential directions proffered by 'feed' versus 'seed' technologies?

Discussion:

Feed technology is clearly preferable to seed technologies if we are primarily concerned with preventing illicit uses. Central control and monitoring of 'precursors' is already a strategy in preventing the use of genetic precursors to pathogens, so that rogue states or individuals cannot easily make biological weapons using 'synthetic biology' technologies. A 'feed' infrastructure might also be able to manage some adapted form of current IP for a nanoware future, serving as the basis for some sort of rights management platform. The source of the greatest promise of nanowares is also the source of the greatest angst, specifically loss of control over a very powerful technology. The history of industry has so far been tied up with the need to accumulate capital before having access to the means of production, much less supporting some sort of distribution infrastructure.

In media industries, lack of access to tools that enable rapid production of copies cheaply, or distribution of records or CDs, dissolved as a barrier to widespread copying as CD burning and internet distribution of music became technologically possible and cheap.

One approach to the P2P 'crisis' posed to contain media piracy has been attempts at greater centralization and monitoring. Internet providers may be pressured by states and media producers to monitor and contain piracy by technological means, such as by limiting upload speeds, or choking off those who download or upload too much. Digital Rights Management (DRM) has also been included at times in the media themselves (although every attempt has ultimately been defeated by hackers). As in the novel, containment of the technology is aimed at preserving the *status quo*, preventing both illicit and positive potential uses of the technology in the process. Extending the current IP paradigm to the technology in *The Diamond Age* would necessarily mean adopting a 'feed' model. Only through some sort of centralized monitoring or control of materials and designs could developers of new products ensure that they are paid for each and every instance of a good reproduced at authorized 'matter compilers'. 'Seed' technology would be the nanoware equivalent of P2P networks for physical goods. Armed with seed technology and a fully compatible design specification, anyone with access to the technology could reproduce an illicit copy of a good at any time, and anywhere. Moreover, seed technologies could not themselves be contained, presumably, once developed as 'seeds' could generate 'seeds' without any possibility of monitoring or control by either states or individuals.

But feed technologies pose the burden of maintaining regulatory control, preventing dissemination of potential, rival technologies (like seeds), and ensuring that those who enter the 'feed' markets will enter with acceptable uses, and become properly remunerated. Nanotech rights management schemes could be developed, by which legitimate copies of goods are distinguishable from illicit copies (though likely hackers would defeat such NRM schemes eventually). As with current IP systems, the state involvement,

Case Study: *The Diamond Age* (continued)

bureaucratic and legal institutions, and enforcement mechanisms necessary to maintain the system would likely be constantly tested by those who seek to undermine or defeat it. Nonetheless, if it could be maintained, it would help ensure the sort of monopolistic control and profit currently expected by innovators and creators who seek IP protection.

The Diamond Age offers us a glimpse of two potential paths both for the technology and for the regulatory and institutional responses to developing nanowares. As we will see through the course of this book, the link between the technologies and the institutions is strong, and our choices about one will affect our abilities within the other. Currently, proponents of strong IP must be wary of 'seed' technologies, because they pose the same risks as P2P has for digital media. Those who are concerned also about the safety, security, or risks posed by nanowares should also choose 'feed' over 'seed'. But are the promises of seed, mainly a low-cost infrastructure, ubiquitous goods, complete decentralization of production, and a true potential end of scarcity, greater than the perceived risks? Can seed support innovation, profitability, and economic growth as well?

The Diamond Age does not offer us a picture of the post-seed world, and so we must try to envision some way in which the risks posed by the technology do not translate into harms, either economic or physical. Throughout the rest of this book we will consider the theoretical foundations for such a world, and examine further cases to try to develop a picture of what real-world mechanisms or institutions might be plausibly conceived to accommodate either future, or even futures unpredicted by fiction.

CHAPTER THREE

The Nano-now

Why do we need nanotechnology? Should we pursue its development simply because we can, or will it serve some set of greater needs, solve current problems, enable both economic and social advancement, or provide other benefits? Will these potential benefits outweigh possible risks? Authors and futurists like Drexler, Kurzweil, Joy, and others I have mentioned tend to agree about the revolutionary nature of the technology, if pursued to its ultimate form, and many scientists, engineers, and even hobbyists are pursuing what could be considered to be intermediate forms right now, with varying degrees of success, and with equally revolutionary implications. Importantly for the subject of this work, those intermediate forms implicate issues we have begun to explore here about the impact of intellectual property (IP), and the need to pursue alternate means of protection and reward outside traditional IP law. These researchers recognize that nanowares will eventually eliminate scarcity, fundamentally alter our economies, and enable innovation, fabrication, and entrepreneurial activity to be decentralized, liberating creativity in ways previously impossible.

Consider the difficulty just fifty to a hundred years ago encountered by someone who wished to bring an idea to the marketplace, much less earn back the cost of his or her investment. Skipping the part about first protecting one's idea through a costly and potentially lengthy patent filing, bringing most products to market meant accumulating capital, or working out a licensing or other contractual deal with a manufacturer who already had, or had access to, the infrastructure necessary to make the product. The difficulty and expense of moving any new item from idea to marketplace ensured, for much of the industrial era, that only organizations large enough, sufficiently capitalized, and capable of withstanding likely losses (like corporations) could risk bringing products to widespread markets. While artisans and small producers could produce items for local distribution, even throughout the industrial era, the scale of production, absent the efficiencies gleaned from mass production, ensured that the products could not be competitively priced, and that profit margins would remain slim. Industrialization (which mechanized production) and mass production (which increased and speeded up the numbers of units that could be produced per worker per hour) afforded inventors efficiencies that helped ensure that if their product found a market, then demand could be met, prices lowered competitively (assuming no patent protection, which allows for monopolistic pricing for the term of the patent), and broad markets reached assuming the availability of some distribution chain.

Ideas are free. Turning ideas into something profitable has been, historically, risky and expensive. Innovation thus became the province mostly of well-financed geniuses, or large corporations, and often both working in tandem. Patents have certainly helped make certain investments less risky in the sense that risks associated with prototyping, testing, then mass-producing inventions could be taken as long as patents had been filed, alleviating some of the risks of other competitors entering the market first. The risks associated with capitalizing the process were offset by the monopoly enjoyed through the patent, allowing prices to be set so as to offset the additional expenses. Innovators who lacked access to capital took big chances when approaching potential financers, or corporations who might agree to license their inventions, especially if they had not already begun to get a patent. Finally, even holding a patent was no guarantee of breaking even, much less not going bankrupt. For an individual inventor, a single patent is just a gamble, a hedge against the possible 'theft' of one's idea and its successful implementation by someone else. But holding the patent does nothing to create a market, nor to induce demand, nor even if there is demand does it ensure that demand will be sufficient to make a profit. Most patents never earn their holders any money. But corporations are able to hedge their risks by innovating in multiple ways at once, often holding patent portfolios with multiple potentially profitable innovations, assuming that many will fail, but that the few that succeed will pay for the losses on others.

For individual innovators who wish to enter the marketplace, the risks can be reduced only by paying up front for a patent, in the hope that the filing and attorney fees involved will be offset by either licensing fees earned by licensing the patent to a company that can afford to capitalize the manufacture and distribution of the product, or that by holding the patent, and marketing the product successfully to potential investors, enough start-up money can be raised to capitalize manufacture and distribution oneself. This is the system that has evolved, and it means there is still a significant barrier to entry for inventors, or designers of new products, no matter how useful and potentially profitable they may be.

This process has been short-circuited, however, in one area of innovation: software. Two developments have made it possible for software products to enter the market with virtually no capital, besides the labor and creativity involved in coding the product. With the advent of software and the internet, production *and* distribution of one sort of product could finally be achieved with little or no capital besides the labor and creativity involved in coding the product. Software has proven to be a boon for some innovators who literally created multimillion dollar products in their homes, using their home computers, and successfully distributed their creations via the internet. Successful computer companies (even a certain hardware company called Apple) have started in people's garages, unlimited by the once-prohibitive costs associated with prototyping, manufacturing infrastructure, and distribution chains. Software liberated innovators from much of the risk (although not the uncertainty) associated with creating something new and hoping the costs of production are recouped. The costs of producing many useful and eventually profitable garage-built software products are frequently only the time and creativity of one or a small team of coders intent on creating the product. The costs of distribution via the internet are insignificant compared to distributing even small physical goods.

Nanowares promise to bring to the world of the physical the ease and low cost of designing, prototyping, creating, and distributing physical goods. They will be the new software, merging the world of information and communication technologies (ICT) with that of material goods. We should look carefully, then, at the impact of IP on innovation in software, and see what lessons it holds for nanotechnology. In fact, many of those who are at the forefront of creating a new infrastructure for a nanotech future have already begun to apply what they consider to be those lessons to new modes of product design, manufacture, and distribution.

A Bridge to the Future: The Trend of 'Micro-manufacturing'

Some of those who have been at both the academic and engineering forefront of nanotechnology are working to deliver some of the benefits it could provide long before actual nano-assemblers or true molecular nanotechnology (MNT) is perfected. A notable example is Neil Gershenfeld, an MIT professor who founded the Center for Bits and Atoms there, and who has long taught a course entitled 'how to make (almost) anything' for master's students. The course developed the core of the idea behind Fab Labs, and informs a general movement to liberate the technology surrounding creating physical goods the way that software and the internet have liberated another, less-tangible mode of creation. The course itself has been made available as 'open courseware'[1] and includes the following modules: 'CAD, CAM and CAE; laser cutting, injection molding, 3-D printing and NC machining; PCB fabrication and layout; actuators and sensors; analog instrumentation; wireless and wired communications. Lecture topics include design tools, NC, waterjet and laser knife cutting, microcontroller programming, circuit design and joining and forming.' This core set of skills and tools is offered as capable of allowing one to actually make (almost) anything. Gershenfeld's book *Fab: The Coming Revolution on Your Desktop*[2] details some of the history, methodology, and potential envisioned by his ongoing project to make the world of atoms as malleable as that of bits. Taking his ideas out of the academy and into the world via 'Fab Labs', Gershenfeld is attempting to make available in a room-sized lab what he hopes one day will be true MNT on a desktop, tied to a computer.

The toolset that is available in a Fab Lab costs just around US$50,000 (a figure which continues falling), and with it one can essentially realize any CAD/Cam-design as a working prototype, assuming the availability of materials. One of the goals of continually developing and refining the Fab Lab toolset, and in placing Fab Labs all over the world, including in many developing countries, is to open up potential routes to innovation, and access to markets, previously foreclosed due to the problems outlined above: capital and risk. Another goal is to eventually refine the toolset such that a Fab Lab could make a Fab Lab. To meet both of these goals and to help achieve these ends, Fab Labs are explicitly un-patented. The tools (including the complete specifications and parts lists) and the software upon which the labs run are open, available to the public for duplication free of charge. The Fab Lab Charter encourages openness, stating:

Secrecy: designs and processes developed in fab labs must remain available for individual use although intellectual property can be protected however you choose

Business: commercial activities can be incubated in fab labs but they must not conflict with open access, they should grow beyond rather than within the lab, and they are expected to benefit the inventors, labs, and networks that contribute to their success.[3]

The near future of Fab Labs includes greater availability of the tools of design and prototyping of new inventions, cheaper toolsets, and hopefully, more entrepreneurial activity. But the IP climate surrounding them might yet hamper their prospects. While users are encouraged to keep their designs open, they may yet seek IP protection. Precluding this would, of course, itself be an affront to openness as it would reach beyond the use of the tools and limit what innovators can do with their ideas. But IP regimes devised in the industrialized West have pervaded many of the countries in which Fab Labs are now available, spread through international treaties requiring developing countries to abide by Western norms. Innovators seeking to create new products, and intending to enter markets, will do so at their own risk just in case their designs infringe someone's IP.

Taking the Fab Lab goals and philosophy to their logical extremes, and the stated goals of many of the field's leaders, will result in replicating replicators, capable of cheaply producing not only prototypes rapidly, but even working parts. Eventually, as with the goal of Fab Labs being capable of being used to create Fab Labs, numerous other nascent technologies that are now under-developed may lead to the fabled convergence, or singularity, not from the nanoscale upwards – but from the meso-scale down. It is worth noting that most of these efforts are currently being developed as open source projects.

The dream of personal desktop fabrication is being pursued at more modest levels than Fab Labs, by open source development projects like RepRap and Fab@Home. These projects have reduced the toolset for cheap, easy, digital fabrication of objects to just one: digital, 3D printing. Three-dimensional (3D) printing is part of the Fab Lab toolset, and involves the creation of 3D objects by building up successive layers of some material, typically a polymer, although other materials are proving to be useful and capable of generating working models and parts, not just mere prototypes. Leapfrogging the marketing of professional tools, such as Hewlett-Packard's recently introduced DesignJet 3D (which retails for around US$13,000), projects like Fab@Home and RepRap have achieved reasonable 3D printer designs for a fraction of HP's desktop 3D printer. The parts for the current Fab@Home Model 2 cost about US$1,600, and a growing number of parts can be 'fabbed' on a Fab@Home.[4] The RepRap Mendel (the second generation RepRap) boasts materials costing only US$520, and has many parts that can be made with a RepRap.[5] The stated goal of each of these and similar projects, as with Fab Labs, is complete self-replication. The implications of success are quite revolutionary, perhaps as much so as fully realized nanotechnology will be.

A New Industrial Revolution?

Even before true MNT becomes a reality, the trends and projects discussed above promise to alter the nature and economics of innovation. The risks of innovation will always remain high, in the sense that very few new artifacts ever really catch on with a large consumer base. But the costs of failure are falling rapidly. Certainly, the possibility of fabricating complex objects anywhere in the world given a complete set of specifications poses threats to current systems of manufacturing and distribution. Decentralized manufacturing of objects means loss of control over the 'first sale' of an object, assuming the object can be completely duplicated. Consider the loss of profits claimed by designers of Gucci or Luis Vuitton handbags due to the proliferation of counterfeits. Sales of music CDs and movie DVDs continue to fall given the cheap and easy ability of anyone with an internet connection to download near-perfect copies. Imagine now the ability to truly reproduce any object in one's home using a desktop device hooked to a computer and the internet. The technology that is the dream of MNT researchers like Neil Gershenfeld will be a nightmare for the owners of IP. The losses currently faced by copyright holders will spread to patent holders as nearly any patentable object becomes easily copyable. Yet it is this same technology that promises to liberate the creative, innovative, and productive processes wielded now mainly by the already well-capitalized. It is this same technology that could alleviate poverty itself, bringing at least material wealth to every corner of the globe, overcoming barriers to ownership, such as costly materials, manufacturing, supply, and distribution chains. One-and-the-same technology is a terrible threat to the established economy, and the ultimate promise for a new one. What will that economy look like, and who will win the inevitable battle?

The advent of true MNT, and perhaps even some mid-range 'micro-manufacturing' phase involving desktop fabrication of objects with multiple materials (in some final, not-merely prototype form, for instance), means a transition from an economy that profits from scarcity to one in which scarcity is no longer an issue. While currently, prices in a free market settle into a state where margins for profit are available even to competing manufacturers in a market for a scarce good (because consumers cannot make the thing themselves, or because the thing is naturally scarce in some region), prices in a world in which anything can be made anywhere at any time (assuming available component materials) will fall inevitably toward zero. If prices for products will only fall, then we should consider the economic problems caused by a replicating-replicator-induced deflationary spiral, and whether such a technology is not only disruptive but also destructive. We need not evaluate models of supply and demand, nor bicker over the accuracy of such models, in a world in which items can be made at will by replicators capable of replicating themselves. Simply put, markets as they are currently construed would cease to exist, at least for things that could be made by such replicators. Perhaps the only surviving markets would be those by which designs or ideas for new things might be exchanged, something like an economy of types rather than tokens.

In a world of near perfect abundance, our economy would be radically different. Without the need for capital, only a few things would remain valued as scarce, including primarily creativity and land. Although anyone with the self-replicating replicator of the true MNT variety might make any object, *designing* new and useful things would likely still only be a skill that some subset of the population would have. Besides creativity, other human qualities necessary for services would remain in demand, as well as high-tech skill-sets associated with maintenance and repair of existing things. Land, too, will remain necessarily scarce as perhaps the last true rivalrous material good. Clearly, as we transition to an economy without scarcity of most if not all material goods, many will find their jobs disappearing, whole sectors of the economy will shut down, and the nature of production and distribution, as well as the nature of consumers and producers in general, will be altered forever. The structure of the last industrial revolution will be irretrievably lost, but what will be gained?

In a world with little to no scarcity, the definition of poverty will change, as will the definition of its opposite. No one will judge one's wealth by the number or quality of possessions, nor will individual possessions (except those, perhaps, with sentimental or historical value) be worth much at all. Rather the worth or value of objects will be only that value each user or owner affixes to the object, even where no more *market* value might exist. Let's assume such a future will come, and that the political and social hurdles that stand in its way are somehow bypassed or overcome. Many will question whether and how innovators will be rewarded for their inventions, or what incentive will remain for anyone to invent anything at all. Perhaps IP laws can be adapted to such a future in order to encourage continued innovation, and to give some reward to those who will still hold the (relatively) scarce, and thus valuable, trait called *creativity*. Can the legal institutions of patent and copyright, which many argue helped propel a fair amount of the innovation and wealth creation of the industrial and information revolutions, be adapted to the nanoware future, or even its intermediary forms involving some form of self-replicating replicators? How would this work? We can see how it will work by looking at the impact of IP on software and ICT, in which scarcity is not a driving economic force, and in which IP laws have been applied to varying degrees of success to create artificial scarcity.

Software and IP, a (Failed) Experiment in Inducing Scarcity?

As we have seen above, ICT and software have proven to be problematic for the application of traditional IP regimes. In many ways, both patent and copyright have failed either to achieve their principle aims of encouraging innovation or to successfully protect and reward authors or inventors. The failures have been both pragmatic and theoretical, and in the process, serious flaws in the ontology of IP have been revealed, as we have discussed in previous chapters and in more depth later in this book. The theoretical failures include conceiving of software as belonging to two previously mutually exclusive categories (copyright and patent); confusing algorithms with ideas and vice versa; inaccurate application of the categories 'abstract' and 'concrete'; overlap of the category idea with that of abstract; and misunderstanding of the natures of ideas and expressions. Practical limitations have been revealed in the failures of either copyright or patent to do what they are claimed to do: promote innovation, and reward authors and inventors for a limited time, after which the public is rewarded by the invention or work of authorship's devolving to the public domain.

Practically speaking, neither patents nor copyrights tend to reward innovators, although technically that is what they were designed to do. Authors have only very rarely been the true beneficiaries of the patent and copyright systems historically, largely because of the mechanisms involved with publishing. The same has been true, more or less, for software. It is not necessarily the institutions themselves that are to blame (although as we'll see, there's plenty to blame on them), but rather the economics of seeking and enforcing protection with IP. Simply put, IP protection of the sort that matters is only really available for those with the capital to finance potential litigation, as well as the will to pursue their claims. Record companies, publishing houses, and large manufacturers can afford to both seek and enforce IP protection, and authors and inventors with little access to capital (before they 'hit it big', for instance, with some work of authorship or invention) might be able to afford to file for a patent, can afford certainly to copyright their works (as copyrights are free), but likely cannot afford to fight any litigation that might result due to someone's infringement or their own alleged infringement (or claims by others that their works are infringing). The history of IP litigation is replete with instances of the little guy getting steamrolled by the big guy. Chances are, if you are not rich, and you are an inventor or author who is challenged by a large publisher, designer, manufacturer, etc., then you will not have the fortitude to do what is necessary to win (even if legally you are in the right) unless you are willing to spend a great deal of money, take a great deal of time, and risk losing in court. The economics of litigation favors settlements forced by those who hold the leverage of capital, at the expense of justice. It is a rare person who has been wronged who can afford to pursue justice despite the cost and risks, on a matter of principle.

Authors and inventors who seek copyrights and patents are thus best advised to do so for leverage in some sort of sale of their work or the rights to their work. Publishing contracts (typically) for music and books favor the publishers first, who, after all, take the risk of expending their capital to make sure the works see the light of day. Royalties for authors who are not already established are scant, and all but the most token advances, rare. A patent is valuable for the inventor, if he or she works out a decent licensing deal (with the help of an expensive lawyer) and may make inventors rich, but the risks of the marketplace ensure that very few ever do become rich, while corporations more easily pool the risks of multiple patents in the hopes that one or two will generate millions. The economic gamble for the small inventor, or unknown artist, is generally too great to take on the established publishing or manufacturing community, and the potential for reward too small to make it worthwhile.[6] Patents are expensive to get, and very expensive to enforce, and so they too end up becoming the property of those who have more capital, less to lose, and the means to enforce them. While there are exceptional cases of garage inventors making it on their own, or garage bands becoming wealthy with 'indie' label contracts, these are exceptions to the rule. With the advent of ICT, specifically the internet, the economics of music publishing (and software as well) has changed irrevocably. Now, not only is it hard for laypersons with little capital to seek or enforce IP protection, but even those with sufficient capital now routinely fail to be able to contain the 'theft' of their IP.

Software was to be the ultimate medium for do-it-yourself inventors, opening up the means of production to anyone with creativity and a computer. Yet today, it remains the large content providers who routinely benefit from IP, whether its movies, music, or software. While the occasional Doom™ might come along as evidence of the ability of basement coders to make it big, the overwhelming number of dollars spent on software goes to well-financed software publishers who either generate code in-house or buy the IP of smaller companies or individuals. Now, given the rise in the popularity of P2P sharing of files, movies, music, and software are shared all over the world, reducing (theoretically) the number of first-sales that would ordinarily generate royalties due to IP. The risk is rising for those who could afford IP, and the rewards are falling.

The battle has also often devolved into technological warfare, as IP owners have sought at various times to prevent the 'theft' of their IP with technology. DRM, or digital rights management software and hardware, has been employed in various ways to prevent copying through technological means. Each time, however, similarly innovative hackers have devised mechanisms to thwart DRM. Some DRM has even allegedly harmed users' hardware, decalibrating hard drives or DVD drives. Money spent on new technologies to stop copying becomes cheaply and easily defeated within shorter periods of time, and users who buy pieces of hardware (like a computer, video game console, or DVD player) assert their right to use the devices they own, and modify them as well, in any manner they see fit. The DRM arms race is an ever downward, tightening spiral, or game of cat and mouse, and the mice continue to win.[7]

IP has failed in ICT both theoretically (IP, as it exists, cannot fit ICT into its categories and principles neatly) and practically. It does not reward nor protect as it should, and the rate of practical failure appears to be increasing. But what of the other stated goal of IP, to encourage innovation? Despite the failures addressed above, is IP still working as sufficient incentive (perhaps due to poor gambling heuristics, or poor understanding of the current economics of inventing and authoring), such that innovation continues unabated, or is there a risk that innovation will slow in the face of IP's failure? Or have other mechanisms continued to fuel innovation even as IP fails?

Alternatives to IP and Innovation in ICT

As traditional IP has slowly failed in the ways described above, new forms of innovation and protection of IP have emerged. Some of these owe their origins to 'open standards' manufacturing projects dating back to the early 1900s, and others arose out of modern efforts, primarily in software development, to stave off the perceived deleterious effects of copyright and patent law on software coding.

Open source methods of innovation reveal the product and process from the outset to a community of developers (and sometimes users) in order to defeat certain negative effects of IP monopolies. The general notion is that the details of the product or process, when revealed to the community, encourage its use and improvement over time. Absent IP protection, users and developers of open source products need not fear any legal repercussions to improving the product or process, and over time a virtuous circle of use and improvement perfects the product or process to the benefit of all. In the early days of automobile manufacturing, a pseudo-open-source approach was adopted by the Motor Vehicle Manufacturers' Association which had created a 'patent pool' on the products and processes involved in members' automobiles. Members of the association could freely use the patents of other members without fear of litigation, allowing for each to innovate freely, freeing capital that might have been required to be set aside in case of litigation, and encouraging the improvement of parts and methods employed in their vehicles rather than paying lawyers to fight to hold and enforce patents.[8]

Open standards have been used for even longer in science and engineering. Units of measurement, standards of scientific and mathematical notation, and other community-based mechanisms that encouraged the spread of information without affixing ownership rights, all are progenitors of the modern open source 'movement'. Open source approaches to innovation, as well as to its precursor science's methods, have helped ensure that communities of researchers and developers can share their forward movement, collectively grow thereby, and even successfully commercialize products as a result without the necessity of monopolies 'far upstream'. The notion of 'upstream' and 'downstream' relates to the level at which a monopoly right might attach. Consider an elevator. The law of gravity is upstream while the particular mechanisms employed to use gravity to move an elevator up and down, in the particular configuration or a certain elevator brand, is downstream. There is a gradient in between the laws of nature and the exact mechanisms employed that is mid-stream, with certain features falling further upstream than others. For instance, the use of cables to attach a box to pulleys is probably further upstream than the particular ratio, the diameters of the pulleys, etc., used in any single elevator design. In many industries it makes sense to not seek patents too far upstream, and to rely either on open standards or patent pools. We see examples now in the development of open standards for video codecs (the algorithms by which video is digitized and compressed) for DVDs and internet. The same was true for videotape recorders and players. By keeping the standards for the encoding and replaying of data open, manufacturers of tape decks, and now of software that records or plays back video, create a virtuous circle by which their machines or software can interact with each other's, improving trust and increasing usefulness so that consumers feel free to buy the product. The use of proprietary standards can have the opposite effect.

The videotape standards war between Beta and VHS offers some lessons in the value of opening up standards, or at least liberal policies of cross-licensing. While Sony was slow to cross-license its video encoding standard to other manufacturers, JVC did so rapidly, increasing the number of VHS recorders and players on the market, and driving demand upward as the price of tapes fell. Users could risk purchasing any one of the VHS brands because their tape collections would play on any other manufacturer's VHS player, whereas a user who spent on Sony's Betamax would be forever stuck to a Sony machine. At first blush, open standards might seem like a poor way to ensure profits. If Sony opens up its standards, or cross-licenses them liberally, it is only asking for competition, rather than locking in guaranteed profits due to an IP monopoly. But by locking up IP so far upstream, an inventor takes a huge gamble that the consumer base trusts them sufficiently to reduce the number of available options downstream. With all but the most respected or economically dominant manufacturers, this gamble is not worth the risk. Potential customers faced with locking in their choices so dramatically will choose another option, unless the product is valued for some other reason, or because brand loyalty is high enough to warrant giving up choices in the future. At the inception of a technology, especially, keeping standards open has proven to be a successful strategy. Only in the rarest of cases have innovators benefited by proprietary standards that remain closed due to IP where the technology monopolized is far upstream.

Open source or open standards dominated the early development of software. Early coders experimenting with writing software on shared servers at places like MIT and Stanford shared their code. 'Hacker' culture, out of which many of the leaders in the world of computers and software came, valued coding for its artistry and ingenuity, and improved upon each other's products. Closed systems, not amenable to tinkering, were frowned upon. The machines were new, at least to those who later founded Microsoft,[9] Apple, and other computer powerhouses, and freely tinkering with code, which was itself freely shared, proved the foundation upon which knowledge, skill, and innovation in software first began to spread. As Steven Levy details in his history of the early ICT era, *Hackers*, openness is considered a virtue, not just a practical necessity for innovation (at least far upstream). The heroes of the early days of software, who liberated the technology for everyday users and put computers in all our homes, turned at some point. Bill Gates stopped believing in the virtues of open source when he bought the kernel of what would become MS DOS, and modified it, working out licensing deals to put it in millions of new computers sold. It was protected then by copyrights and patents. Steve Jobs, who helped finance the early days of Apple by selling 'blue boxes' which could be used to hack phones to make long distance calls for free, became an IP believer when Apple took off. Current systems like OSX and the iPhone operating system are open to developers who wish to make code to run on top of them, but the code itself is not open to all.

Reacting to the commercialization of software, when licenses for some high-end products could range from the hundreds of USD to the tens of thousands, hackers began to develop their own, free products and built new licenses to guarantee they would remain free. Richard Stallman was an early pioneer of creating free software equivalents of expensive protected software. Unix was one such expensive system. The GNU operating system (which recursively stood for 'Gnu is Not Unix') was a free alternative to Unix, but it came with a minor catch. To use GNU, you had to agree to the terms of the GNU license, called the GNU GPL, for GNU General Public License. Otherwise, GNU was free, and meant to be wholly compatible with its commercial cousin, Unix. GNU was built on the principles enunciated by Richard Stallman at MIT, and the GNU project was begun in 1983 and continues today. Stallman first quit his job at MIT in order to prevent MIT from claiming any IP rights to the project, and later published the *GNU Manifesto*, defining the goals and purposes of the free software movement.[10]

The GNU suite of software became widely known and used, especially on college campuses, in computer labs, and among hobbyists in the mid-1980s. It encapsulated the philosophy that code ought not to be owned, nor that the source code could or should be hidden from users (a compiled program you might buy at a store excludes the source code, so you cannot find out how it works, nor tinker with it to make it better). The Free Software movement, more or less begun by Stallman who remains a major influence in it today, is motivated by the 'hacker' ethic, according to which the things we make and use ought to be transparent to users, and available to tinkerers to improve upon and modify. As the GNU project evolved and eventually developed an operating system, word processing software, and numerous other applications, Stallman developed the first GNU General Public License, or GNU GPL. The GNU GPL is meant to ensure that the software distributed under it remains free, and that no one profits by reselling it, or by attempting to monopolize it through traditional IP law.

One might very well ask what business sense 'free software' makes. How can one *profit* from making things freely available at no charge? The free software philosophy argues that the community of users profits simply by having better software available, freely, and openly so that anyone can improve it. Indeed, recent surveys show that free software pervades business operations throughout the world, and large segments of the information backbone of large corporations, and the internet itself, run on free software. Tools like Apache (a major platform used for webservers), Linux, Mozilla

Firefox, Sun's OpenOffice (a complete competitor to Microsoft's Office suite), and numerous other applications are now available under 'open source' licenses.[11] Open source is an offshoot of the free software movement, and has adopted many of the core institutional tools of the GNU GPL, but has also developed business credibility, and a devoted user base which pays, often, in various ways to use these otherwise free (and still open) tools. The movement toward open source has helped both legitimize and commercialize, yet keep software coding open. Because of the commercialization and adoption by large corporate entities like Sun Microsystems, Stallman has publicly opposed open source and continues to advocate free software. Nonetheless, a growing number of software projects, at every level of complexity and with varying degrees of commercial success, are now committed to calling themselves open source projects.

Importantly for our purposes, open source has begun to move out of the realm of software and computers into methodologies for delivering other goods as well. A critical difference between the free software paradigm and open source is that under the open source philosophy, people are entitled to profit from open source products, if only in limited ways. For instance, companies like Red Hat, which offered services surrounding its version of Linux (an open source operating system), built its business model on the provision of services and products *on top* of the free operating system it sold (Red Hat has since merged with another Linux-based system called Fedora). Sun Microsystems, as mentioned above, has a complete competitor to Microsoft's Office™ suite called OpenOffice. Hewlett Packard has been achieving some success by offering engagements with companies building upon the installation of open source software. The margins they are realizing, and the savings that the companies they contract with achieve, are part of their value statements.[12] So how can one profit on top of a product that is free? By changing the nature of the transaction – by charging not for the product, but for some *service*, or by selling some proprietary version apart from a free version.

Open source projects have shifted the conception of value away from goods to services, and have proven to support the proposition that creating products out in the open, without worrying about locking up the IP, can spur both innovation *and* profits if done right. Open source is also the model behind some of the home-grown fabrication projects discussed earlier. Can open source spur both innovation and profitability for nanowares in this transitional mode toward MNT, and how would an economy flourish and grow in such a transitional phase, or in a true MNT future if open source prevailed?

Innovation and Growth: Profiting without Scarcity

Why do people innovate? Why do certain people wake up one morning and decide to create something new and useful, or to delve into some hitherto unknown aspect of nature to figure out why some physical phenomenon happens and how? Why do some people choose to make up stories, or songs, or make paintings or sculptures? One theory is that all of these activities occur because all of these creative impulses lead to profit. Part of the profit *now* achieved from each of these activities and products now stems from the monopoly rights afforded by IP regimes. Proponents of existing, traditional IP regimes argue that without them, the rate of innovation and growth experienced in the past 150 years would not have been possible. Let's assume this is the case, just for the sake of argument. Perhaps because of the nature of industry and the innovations that drove it since the industrial revolution, monopoly rights helped to buffer the risk of developing infrastructures for both manufacturing and delivery of new products, and thus encouraged large scale, expensive capitalization of new discoveries and inventions as eventually profitable consumer technologies. Although there is nothing but correlative evidence to support this hypothesis, and no *causal* necessity to it, and many economists disagree regarding its historical accuracy, let's assume for the sake of the rest of this argument that it is the case. The question remains: Even if IP was necessary for innovation and growth during the industrial age, is it necessary now, and will it be necessary for the *next* postindustrial age in which nanowares predominate?

Let's examine the possible reasons why IP might have been necessary to growth and innovation for the past 150 years and extrapolate. One argument for IP's necessity has been expressed above: the cost of investment in the research and infrastructure for the development, manufacture, and distribution of a new product *requires* some legal protection by way of a guaranteed monopoly for some period. The monopoly rights afforded by copyrights and patents ensure that, should a product or work of authorship prove to be in demand, then the public benefit conferred by their provision is rewarded by some term in which no one else may market the work or receive profits from first sales. In return, the monopoly right is set to lapse at some point, but in the near term, authors and investors are encouraged to take the risk, expend the costs associated with the design, production, and distribution of the work, and so innovation is encouraged and invention rewarded.

The manner in which works have traditionally been produced, both aesthetic expressions (the subject of copyright) and utilitarian ones (the subject of patent), has perhaps warranted the assumption that IP is necessary to encourage risk involved in innovating. After all, publishing a book, making a movie, designing a computer, or any new device, all required a great deal of money, especially if one wished to not only create the object, but place it into a crowded stream of commerce. Part of the reason for the expenses involved has been scarcity, which, as we have discussed a bit already, helps drive the prices of goods in a market. These forces are the familiar ones involved in the law of supply and demand. Consider books: while books were once hand-lettered by monks (a service that was itself rare producing even rarer products) the advent of the printing press made the service of book printing and thus the product less rare. Prices fell. Over time, the cost of printing books fell, driven in part by computerization in the latter half of the twentieth century. But physical book prices and the costs of setting up the infrastructure for printing and distributing books in profitable numbers are still driven by scarcity, even as the prices have fallen, and the margins have increased due to technologies like 'print on demand'. Printing a book still requires a press of some sort (even if it is a machine that can print and bind complete books on demand). Books also require paper. These physical necessities are necessarily scarce, as all physical goods are. Even replaceable goods like food and paper, which can be grown in theoretically limitless quantities, will always remain scarce because at any one time there is only a limited amount of the good available for use. The same is true for all traditional modes of production of both aesthetic and utilitarian goods. Beyond the cost or scarcity of the creativity involved in the design of some good, there are always some scarce products involved in the manufacture and delivery of the final product into the stream of commerce.

ICT changed all of this with the convergence of software and the internet – the perfect, frictionless delivery device for theoretically never-scarce goods. Software now suffers almost none of the scarcity issues of other industries, and essentially the only cost involved in the production of a new product is the creativity and time involved in coding. As we have seen with successful open source products, even these 'costs' can be provided essentially 'for free'. Innovators in ICT recognized rather early on the value of openness upstream, and even downstream, in promoting innovation where capitalization is not an issue. With ICT goods, research, development, and distribution no longer created expenses that needed to be recouped through monopoly rents. As we have seen above, even in the manufacturing of physical, scarce goods, forgoing IP upstream is often a rational choice for innovators and entrepreneurs who seek market penetration without the costs associated with IP disputes or potential disputes. For ICT innovators especially, given the decrease (ordinarily) of scarcity and thus costs of capitalization for design, development, and distribution, open source solutions have often proven to make sense. They reduce the costs, and lubricate the development cycle, providing input by a community which also aids in distributing the work. They also build consumer trust and loyalty.

The fact is that innovation still occurs in a climate in which free alternatives to commercial products compete equally. The fact also is that the industry remains profitable. The software industry is thriving, even as other media and manufacturing industries suffer. The gross profit margins in the software sector most recently ranged around 75 percent.[13] This is a significant margin, partly due to the fact that the amount of capitalization required in this sector is lower than in hardware industries, as discussed above. These margins, and the fact that more companies are realizing the value of their sales can be acquired in *services* as much as in products, continue to argue the case for providing at least some of the sector's new products and innovations in software under an open source model. For the same reasons, innovating in nanowares might similarly argue for and prosper from an open source model.

With MNT, and perhaps as transitional forms of nanowares become perfected, scarcity will continue to decline in importance. Without scarcity controlling the provisioning of materials, infrastructure, or distribution, marginal costs associated with design, development, and distribution will similarly fall. At some point, as true MNT is perfected, the assembly of any item at any place in the world will be as cheap and easy as downloading a piece of software. The costs associated with the invention of a thing will depend upon the *value of the creativity* involved, and little more. Of course, creativity will remain scarce, as will services associated with the design and distribution of *types*, rather than tokens. It seems likely that those involved in the creation of the transitional forms of nanotech described earlier, and who remain active in promoting the science involved in true MNT, recognize that the goal of ending scarcity as a physical impediment to well-being requires open source development at the inception of the technology. Fab Labs, RepRap, Fab@Home, and similar projects are all currently open source. As opposed to HP's 3D printer, which is proprietary and relatively costly, the open source alternatives being developed are accessible for general consumption. They are also open and subject to improvement by the community of users. Competition among the various projects is continually pushing costs down and improving the systems available. People are even making money with these systems, as they do in other open source industries, by selling their services in building open source fabricators for those who prefer not to buy them used, or to make them themselves. What might the economy of MNT look like? We might be catching glimpses of its form by examining the current landscape in the ICT sector, and watching micro-fabrication evolve as it has.

Capitalism without Capital

One direction those who are pursuing MNT might pursue is the path of the *status quo*. Developers could well file for patents on their products, and pursue stronger enforcement so that physical goods don't become the next heavily pirated object on peer-to-peer networks. This path will be expensive for developers, ensuring that only large corporations who typically can take the risks associated with patents and their costs will dominate the innovative landscape of this new technology too. It seems likely that many innovators pursuing these technologies have already soundly rejected that path, judging by the proliferation of open source, grassroots development projects. Artificial scarcity by means of IP would clearly clog the innovative pipes upstream, and slow the promise and progress of these technologies early on. Some might fear that without strong IP, and with free copying, sharing, and improvement of individual innovator's ideas, capitalism as we know it will cease. Capitalism is built, after all, on the specialization of trades, scarcity of goods, and the invisible hand of supply and demand directing rational pricing in free markets. Without scarcity, prices would seem to always trend toward zero, and thus profit margins would likely cease to exist.

But this is likely not going to happen if we can fairly extrapolate from the model of ICT. Specialization will continue at the creative stage of product design. Improvements to existing designs require human ingenuity, and services surrounding the provision of software will be similarly important (and maybe even more so) with physical goods and their complicated relations among moving parts. People will still pay others for doing things they cannot themselves do, even where they need not pay for no-longer scarce physical objects. New, improved objects will always be in demand, and those who can design or improve things will be similarly in demand. Finally, the need for profits will slowly come to be reduced, even as profit margins continue to climb (because of the low costs of nonscarce goods). Innovators are also consumers, and as prices for goods fall, so too will the need to accumulate capital.

The transition will no doubt be difficult, and economists might rightly worry about the deflationary effect the growing lack of scarcity would have on individual and world economies. Some things, however, will always remain scarce, including both land and creativity. Creativity will be the fuel of the nanoware economy, and educating people in the creative and productive uses of both transitional forms and true MNT will be highly valued. The services that we provide each other will similarly remain valued, and valuable. Perhaps food will be replicable via MNT, and some are seeking ways to use synthetic biology as a shortcut to alleviating scarcity of food and energy, but during the transitional phase as we move toward real MNT, land, food, and the services associated with providing basic needs will remain scarce. Of course, the end of scarcity itself remains the noble goal of converging technologies, and the scenarios and arguments made here are only outlines. But we can see in ICT, and in the early stages of the nano-now, that this potentially disruptive technology can be both profitable and revolutionary at once, if we embrace the best in our creative abilities, and learn lessons from our recent past.

Case Study

Nanowares in the Market

Many of those who are actively pursuing a nanotech future are focusing on achievable results that will presumably aid the development of a true MNT infrastructure. Specifically, the ability to construct identical copies of the same object, not at the nanoscale, but at a useable 'meso' scale, delivers some of the same promise of MNT. This is why this book considers such 'micro-manufacturing' to belong on a spectrum called *nanowares*. It is a transitional phase, and it is happening now.

Mass production has been the model for corporate success in the marketplace of physical goods. In order to produce and distribute goods at a scale that can produce profit margins deemed to be sufficient, factory-driven mass production has been more or less universally adopted as the best model. While this model has arguably worked well for consumer goods whose market demand is great, it means necessarily that products which enjoy only a niche market have had to seek smaller scale manufacturing solutions. Until recently, such solutions were hard to come by. In the past decade, however, it has become possible to have niche manufacturing done by contract, especially by competitively priced Chinese manufacturers.

But this model too is vulnerable, and costly, and does not solve the problem of distribution infrastructures. Cheaply made, small lots of factory-fabricated goods made in China still need to be shipped, either back to the designer who ordered them or to the eventual customers. One solution is to take yourself, as the designer, out of the supply chain, and to make the goods where they are wanted – by the customer, essentially. This is the ultimate goal, of course, but it will be a while before RepRap or Fab@Home bots are able to accomplish this technically. Meanwhile, some are attempting to enter the market with some of the institutional infrastructures they think will underlie the coming home-fabrication revolution. Two concurrent trends, sometimes intertwined, are precipitating the nanoware future: micro-manufacturing (by which we mean manufacturing on a smaller scale than that done with traditional factories, and with custom-sized runs) and distributed or decentralized production.

While Fab Labs are enabling the easier, cheaper design and prototyping of products, micro-manufacturing and distributed production allow for bypassing problems associated with capitalizing

Case Study: *Nanowares in the Market* (continued)

large production runs, warehousing large lots of goods, and cheapening the supply chain. Reaching markets is another story, and companies that are producing small lots of goods are being aided by websites like http://www.etsy.com. Etsy enables producers of goods to find customers who are specifically interested in custom-made, one-off products. Customers can also custom-order goods made to their specifications from producers willing to tailor-make them (http://www.etsy.com; another similar site is http://ponoko.com). Most of the items available are handmade by artisans, but the success of this marketplace suggests that manufacturers of limited-run items can find alternative markets without significant capitalization. eBay also stands as an example of a marketplace that serves the emerging trend of micro-manufacturing. The costs of entry to internet-based, mail-order marketplaces are low, and margins for products can be increased without increasing unnecessarily the costs of the products themselves. Even large companies are choosing micro-manufacturing as an alternative to large production runs and warehousing, and a number of infrastructural, trade-oriented solutions are becoming available. Micro-manufacturing online is a major portal to products, associations, and a network of those involved in this emerging trend (http://www.micromanu.com/).

Another approach is to distribute the type rather than the tokens, and companies like Arduino are doing this, in a sense. While Arduino (http://www.arduino.cc/) manufactures and distributes copies of the Arduino microcontroller (it manufactures them at its plant near Milan, Italy, using a small, 'pick and place' micro-controlled robot), it also distributes the complete set of instructions for making its micro-controllers, billing its product on its site as 'Arduino is an open-source electronics prototyping platform based on flexible, easy-to-use hardware and software. It's intended for artists, designers, hobbyists, and anyone interested in creating interactive objects or environments'. Arduino is consciously financing with its profits the total distribution of the manufacturing infrastructure. It is also creating an institutional infrastructure, having adopted a form of the Creative Commons license for use in the distribution of the complete set of manufacturing instructions used to reproduce its controllers. Everything it makes is open source, and hobbyists, serious inventors, and others who envision the total decentralization of production are adopting its tools and similar models to enter the market cheaply, without any supply chain at all. Even while anyone can download the plans and make the final product themselves, and even improve upon it, Arduino profits from orders of completed boards, or parts that can be ordered by those who wish to assemble the product themselves.

Another approach is being taken by Bug Labs, which sells open-source based modules, a sort of preassembled, hardware app toolset, that can be assembled into new devices according to the consumer's desires (http://buglabs.net). Even while the Bug Labs parts are all based upon open source components, the company can profit by selling preassembled parts. A community of users share their 'bug' based inventions, and more people order the parts, driving a virtuous cycle.

These various efforts at creating new technological precursors to true MNT fit firmly within our inclusive category of *nanowares*, and the institutional approaches being undertaken to accommodate them are indicative of the inadequacy of current institutional approaches to the merger of hardware and software. It is not surprising that a prevalent model being used to both encourage and protect emerging nanowares is an open source one. These efforts are geared toward devising means to distribute, ultimately, not tokens but types. Ideally, eliminating the supply chain necessary for a final innovation will be one of the critical achievements of nanowares. IP law as we know it cannot protect this sort of exchange, nor is it clear that it would encourage innovation, although it might best ensure monopolistic ability to control prices and assure profits. Patent law prohibits the reproduction of a thing, and copyright protects the reproduction of non-'thing' expressions. Micro-manufacturing solves the problem of capitalizing and warehousing large lots of things, and is a step toward nanowares, but distributed production,

Case Study: *Nanowares in the Market* (continued)

by which the types themselves are reproduced at the point of purchase, is cheaper, more efficient, and arguably truly a nanowares phenomenon.

In the examples above, the ongoing conundrum remains profiting. Arduino distributes production completely when it distributes its full specifications on an open source model, but cannot profit from that distribution via patent, not if they wish to keep the product itself open. The value statement for keeping the ultimate product open is that, by so doing, users have actually improved it. They still also profit from sales of the (now improved) product to those who would rather not construct it themselves, or who purchase the parts from Arduino, and then do the final assembly. So while Arduino's open source platform is earning them profit, they still have not merged profitability into an infrastructure for distributing the types rather than tokens. Bug Labs is reasonable middle-ground, leaving open all of the parts, and distributing them at a profit (even while leaving the parts themselves amenable to copying by being open). Bug Labs's value is in providing the service of delivering an underlying infrastructure for further innovation. They also profit by maintaining a community of users whose growing interest and expertise in improving the parts and infrastructure may tip over into a profitable hardware version of Apple's App Store.

The alternative to each of these approaches would be to develop some sort of licensing scheme that would be modeled after current IP, but that could accommodate the distribution of types, and maintain monopolistic control over the tokens produced at the purchase side, and yet keep the purchaser from violating the prohibition of reproduction. Patent law's current prohibition against reproduction would require each purchaser of a type to get such a license so that the construction of the token would be legal. It seems likely that this approach will be tried as the trends of micro-manufacturing and distributed production become more popular, and their value statements increase. The result would be something like an End User Licensing Agreement (EULA) for hardware, much like those we currently agree to when installing software.

CHAPTER FOUR

Law and Ethics: Rules, Regulations, and Rights in Nanowares

We are interested in promoting beneficial technologies, which nanotechnology certainly promises to be, and doing so in ways that are efficient, innovative, and ethical. Not all regulation is primarily proactive like intellectual property (IP) rules are by encouraging innovation through offering incentives. Some regulation is negative, restricting what may be done initially, and giving no exclusive rights to anyone. Stem cell research, for instance, was deemed not worthy of funding during the George W. Bush administration, when all but a minor subset of federal funding of that research was halted by presidential decree. Governments can spur research with funding, or inhibit it through laws. They can promote investment, and try to encourage innovation through application of IP norms, or by re-imagining those norms and their roles (as I encourage in this book).

So far we have concentrated on the nature of traditional forms of IP, their history and applications, and pragmatic concerns facing the future of these forms if applied to nanotechnology. There are social and individual concerns that warrant regulation in some form, whether by governments, internationally, or by researchers and innovators themselves. Areas in which some form of rule or regulation may be warranted include environmental concerns, safety, and security. Lessons from the introduction of past technologies, and harms that have befallen individuals, populations, and the environment, serve as examples by which nanotechnology can be more carefully introduced into the stream of commerce, and unnecessary harms avoided. We should be mindful, however, that inflation of risks and irrational public fears or aversions have also proven harmful as worthwhile new technologies have been stifled or delayed just when they were needed most.

Regulation poses risks as well. Sometimes safety and security issues are poorly understood, and not appropriately anticipated, while the attention of regulating authorities may be concentrated upon irrelevant features or possibilities. Meanwhile real potential harms that are unanticipated are ignored. Moreover, the costs associated with regulation, both economic and social, may be greater than the risks are actually worth. Overall, regulating too much too early may be wasteful, and yet failing to properly identify real hazards or harms can be disastrous. Weighing these costs against one another is an inexact science, and technology assessment is but a modern, less-than-perfect reaction to past failures, with little evidence that it works better. The ethical case to make for caution is easy, even where the risks are difficult or impossible to properly gauge. The potential for public distrust over an entire sector of technological development based upon harms caused by risks that could have been anticipated or avoided argues for at least a public, concerted effort to confront such possibilities and effectively plan for them. Worse yet, the potential for real public harms weighs heavily against blindly pursuing the technology absent some deliberate caution. Yet the ethical necessity of caution does not mean necessarily adopting what is known as the *precautionary principle*,[1] which led in the recent past to late adoption of important technologies whose harms appear to have been overblown.

Although we have briefly touched on sci-fi, dystopian visions of 'grey goo', and dismissed these as not imminent and perhaps technically not feasible harms to contemplate anytime within the next ten years, there are other potential harms that bear considering, and over which some regulation, or at least self-restraint by researchers, should perhaps be exercised. The risks posed by the technology and its near-relative, synthetic biology include environmental harms, safety hazards, and security risks. Because of the nature of the technology, which as we have discussed is very, very small, research on nanotech will likely become harder to contain, track, or regulate as it matures. Thus, it is now a good time to begin to consider the reasonably anticipated risks and to take what measures can be taken now to avoid some of the more apparent harms. We'll discuss below briefly each of the areas outside of IP law that warrant considering regulation, and examine the potential benefits and harms that regulation (both under- and over-regulation) might reasonably pose.

The Environment

Recently, ethicists and innovators have begun to seriously apply the notion of justice intergenerationally. Borrowing from the legal philosopher John Rawls and Immanuel Kant, a picture of the nature of intergenerational duties requires us to consider the impact of our intentional actions upon future generations, as well as upon our contemporaries. One significant medium of impact for our actions in developing and implementing new technologies is 'the environment'.[2] Past failures argue in favor of extending the scope of our ethical inquiry regarding new technologies. We have spent as societies billions of dollars remediating environmental harms resulting from technologies that were implemented with little concern (and sometimes, little awareness) of their potential impact.

The budding chemical industry of the late nineteenth and early twentieth centuries produced enormous amounts of wealth, and helped bring about a revolution in lifestyles, including the introduction of cheaper, safer, and more effective chemicals and products for hygiene, the production of pharmaceuticals, and widely used household products. It created employment as well, and lifted tens of thousands of workers into the middle class throughout North America and Europe. The legacy, however, of the early days of the chemical industry is hundreds of highly toxic sites, where industries made chemicals or disposed of their waste products without adequate regard for the consequences of their actions over time.[3] In the mid-1960s and early 1970s, governments began the difficult and costly task of finding, logging, and cleaning toxic sites through the world, often only after health effects on local populations had already taken their toll. In the 1940s and 1950s, the emerging nuclear industry exacerbated and complicated the nature of environmental pollution. Nuclear wastes have ever since been accumulating around the world, primarily at nuclear power generation sites, in forms that have proven difficult to contain, and whose effects have taken a similar toll on human, animal, and even plant health.

Waste products have not been the only source of environmental contamination. Sometimes, products themselves later are discovered to pose health and environmental hazards. A prominent case is that of DDT, which was an effective and useful pesticide that helped with the so-called 'green revolution', by which the agricultural industry has been able to keep up with the contradictory demands of a growing population and shrinking amounts of arable land. Unfortunately, DDT resembles dioxin, and was found to be a deadly environmental contaminant, which was killing songbird populations even while it was fighting pests in food crops. Moreover, dioxin was a persistent environmental contaminant that worked its way up the food chain into human systems, causing cancers and other health problems. Asbestos shared a similar story, hailed as a miraculous flame-proof mineral suitable for insulation, and even capable of being woven into clothing; its toxic effects were only later recognized. Because of the size and structure of asbestos particulates, they were found to be carcinogenic when inhaled, and they were all too easily inhaled.

Finally, the example of carbon dioxide poses as the focus of a modern, ongoing dispute regarding the unintended consequences of scientific progress and industrial development. CO_2 is a natural product of respiration by most animals, as well as a natural food source and product of decomposition for plants. CO_2 is necessary for life on earth as we know it. However, its concentrations in the atmosphere have been historically low, measuring in at below 1% of the total concentration of gases (0.033 percent, to be precise). Most of the earth's carbon has been sequestered in benign forms like living plants, minerals, and the products of plant and animal decay. The latter have been sources of fuel for a long time. Wood-fires and then coal proved to be valuable sources of heat in northern climates, and they have been used as household fuels for thousands of years. In the past century and a half, however, with the growing energy demands of industrialization came an increasing demand for 'fossil' fuels. As scientists now generally agree, although fossil fuels helped drive much of the economic and technical progress of the industrial and now post-industrial era, they came with a cost. By releasing significant quantities of otherwise sequestered CO_2 into the atmosphere, we may have changed the earth's climate for some time to come, even if we now successfully begin to transition to other, 'carbon neutral' energy sources or practices. This damage may be largely related to human activities which, while damaging the environment perhaps irreparably (or at least with long-term consequences before reversibility), were the unintended (and largely unknown) consequence of factors that aided us in reaching our current level of development.

The lesson of these past failures is that new technologies carry unanticipated dangers, pose incalculable risks, and often have unintended consequences. These risks and consequences may outweigh the value of the technology, and scientists and innovators alike ought to try to calculate as best they can the risks that can be anticipated, and the chances that unanticipated, unintended consequences may prove harmful, and to what degree. Of course, the nature of unanticipated consequences is that they are opaque, and as in each of these examples, foreseeability is questionable. It may well be that the best anyone can accomplish at any time during the early stages of a technology is to extrapolate from what is known about the fate of similar technologies, and mitigate as much as possible the possibility of irreversible harm.

In fact, because of past miscalculations in other sciences and industries, and unanticipated harms, some of the environmental harms posed by nanotechnology and by its biological cousin 'synthetic biology' are being anticipated and prepared for. Numerous professional, national, and international efforts have already been started to inquire into the potential harms of nanotechnology and synthetic biology, and attempts at regulations regarding environmental emissions of nano-particles have been promulgated. Meanwhile, a balance is being struck between mere caution and the precautionary principle, by which potential harms are avoided by *extreme* caution without regard to the risk (the harm factored with the *chances* of the harm). It seems at these early stages that this balance might avoid the path that Europe took in its approach to genetically modified foods, which were all but banned. The precautionary principle's application in the case of GMOs set countries who applied it back several years technologically, leaving them now to try to quickly catch up. In nanowares, this appears not to be happening. Even so, rules regarding environmental release of nano-particles, and application of nanotechnology to consumer products, are emerging with an eye on the past effects of chemical technologies and biotechnology, as well as the lessons from CO_2 emissions.

Environmental concerns do warrant special consideration, as products that are released into the environment often present persistent harms. Whereas exposures to mechanical technologies, or even testing of pharmaceuticals, pose individual risks to users or subjects.

Releasing products into the 'environment', by which we mean that exposures often become unknown to the exposed, and dispersed geographically, poses threats to individuals who are not necessarily willingly put at risk. If nano-particles emitted either during manufacture or after manufacture and in products enter the environment so that they pose threats to those beyond the ones who knowingly manufacture or consume them, then a greater ethical duty emerges. This duty was overlooked as chemical wastes were disposed of by Hooker Chemical at Love Canal, which became, in the 1970s, the hallmark case of environmental toxins due to chemical manufacturing. People were injured who had no connection with the industry, and who had not directly benefited nor participated in either the manufacture of chemicals or disposal of wastes. These victims were *innocent* in ways that others were not, in that they never gave their consent. As with the Belmont Principles of applied bioethics, we should consider the consent of those affected by our actions when determining the level of care owed.[4] People may be free to choose to submit themselves to risks, but environmental consequences affect others who made no such choice.

Cigarette smoke is composed of nano-particles. These particles have been found to be carcinogenic to a certain segment of the population. People are, of course, free to expose themselves to hazards that are self-directed, as smoking is. But if second-hand smoke, for instance, poses an elevated risk of cancer to non-smokers who can make no choice about inhaling it, then there should be some consideration of the duties owed, and by whom, to those who are environmentally exposed through no choice of their own. CO_2 is persistent, and the accumulation of it over time will take decades to reverse. The same is true of radioactivity. When potential, persistent, and difficult to contain hazards enter the environment, affecting those who are not direct beneficiaries, or who are not end-users, or maybe even not aware of their exposure, an ethical duty arises. Manufacturers and consumers of products that can have negative, persistent environmental consequences have a duty at least to inform, and more properly to mitigate the potential effects to the greatest extent possible. The duty to inform extends to informing both users and non-users who might be exposed, and the duty extends to future generations for whom the exposure may not be willing, nor have any direct benefit (in case the technology falls out of favor, yet the harms persist).

These duties may imply some measure of regulation, but they mostly encourage scientists and innovators to do sufficient testing, and explore the potential for negative environmental impact, before releasing a product into the stream of commerce, and before designing manufacturing processes that may result in releases. Moreover, regulation can be self-originating. In the best cases, scientists and industries are properly cognizant of the long-term impact of their actions, and voluntarily take proper precautions. Overall, this encourages increased public trust, responsible and unimpeded inquiry, and useful innovation with minimal consequences.

Beyond environmental concerns, which are warranted and about which scientists and innovators ought to be always concerned, nanotechnology (as with all technology) poses potential risks to those involved in the research and manufacturing, as well as to knowing end-users. How should those at the forefront of the technology consider these *safety* concerns?

Safety: Ethical Duties in Case of Consent

Humans engage in all sorts of risky behaviors. They often do so perfectly voluntarily, with full knowledge of the potential harms, and even of the chances of those harms. Every day, people choose to drive their cars, which is one of the most risky, regular behaviors one can have. The chances of harm from automobiles, and the severity of those harms, are much higher than the risk of harm due to any existing genetically modified organisms. To what extent must those involved with the development and distribution of technologies inform and protect direct end-users, consumers, researchers, and laborers involved directly with that technology? To what extent ought governments be involved in regulating the duties involved?

A general principle of liberalism as developed in the Enlightenment is that the extent of governmental regulation of private activities ought to be low, or non-existent. Free markets ought to be encouraged in both ideas and economics. The invisible hand of the economy will adjust our behaviors according to the harms or goods that flow from our unimpeded activities. Although this set of propositions remains generally accepted, and the spread of free markets and prosperity have been more or less correlated, there are notable exceptions. These exceptions have been tolerated especially where some 'market failure' has been perceived (even where it might not have actually occurred). 'Market failures' have certainly occurred in the nexus between ethics and technology. We have noted above numerous cases where technology progressed without sufficient knowledge or reflection, helping to cause or contribute to harms that could have been avoided. Sometimes, lapses of judgment in the form of simple negligence were to blame. In other cases, people did what we might consider to be *morally wrong*: introducing a technology known to cause harm without properly alerting users, mitigating damages, nor accepting the responsibility. Sometimes, even when products were developed and marketed without any possibility of foreknowledge regarding possible harms, when evidence of harms was later discovered, no further actions were taken. Taking tobacco as an example of the latter, we can see that sometimes industry fails to self-regulate, and markets similarly fail (in the case of tobacco, partly due to the addictive nature of the product) and so external regulations may be perceived to be necessary.

Why does self-regulation sometimes fail? Because we are morally imperfect, and even the best-intentioned people have their judgments clouded by conflicting duties, self-interest, and even self-delusion on occasion. Market failures in ethics are responsible for the foundation of modern applied bioethics. The current set of foundational principles in bioethics emerged from the Nuremberg trials of Nazi physicians who had, sometimes motivated in part by genuine scientific curiosity, or even by the desire to develop better medical knowledge and techniques, conducted experiments on human subjects that were clearly immoral. The universally repugnant atrocities that emerged in the Nuremberg trials inspired the development of the Nuremberg Code, a set of ethical principles for future research on human subjects. Among these principles are the necessity of voluntary consent, the beneficence of the research (it is aimed toward good ends with good intentions), the requirement to avoid unnecessary harm and to not take unnecessary risks, and that the research should only be done by qualified researchers. These principles derive from common moral theories espoused by philosophers over the past several millennia in some form or another. Yet because lapses by researchers have recurred more regularly than we'd like to think, the Nuremberg Code's principles have come to be formalized in a number of professional codes, as well as institutional rules, regulations, and even laws. Even after the Nazi atrocities, researchers failed to abide by the Nuremberg principles in a number of noteworthy situations. Staley Milgram's psychological experiments at Yale, which were designed to discover the source of unethical behavior (to show why good people will do bad things when ordered to do so by someone in a position of authority), were arguably unethical in light of the Nuremberg principles. In those experiments, Milgram did not give properly informed consent to his subjects who were led to believe they were administering to actors (unknown to the subjects to be actors) electric shocks as punishments. Milgram's experiments were begun in 1961 shortly after the trial of the Nazi war criminal Adolf Eichmann began.[5]

It was only after the famous Tuskegee Syphilis Study that the Nuremberg Code came to be formalized into legally enforceable rules and modern bioethics became a more formal, applied field. This study lasted forty years, and followed a cohort of syphilis-infected sharecroppers in Alabama. They were all African Americans, and they were given regular medical checkups as part of the study, which was designed to discover the full range of symptoms over the course of the disease. But while the study began before any treatments for the disease were known, over the course of the study penicillin was found to be an effective cure. Nonetheless, the subjects of the study were never given access to the drug, and they remained untreated, and uninformed about the existence of an effective treatment, for decades. As a result, the study subjects deteriorated due to untreated syphilis, even while others who were merely victims of the disease, and who did not participate in the study, were cured. The study itself lasted until 1972, and continued following the original cohort, collecting data. The study ended only when the press learned of the plight of the Tuskegee Study subjects and, in the face of public outcry, the study was terminated.[6]

In the wake of the Tuskegee Study, numerous efforts were undertaken to prevent future breakdowns of research ethics involving human subjects. An independent commission was formed which authored the seminal Belmont Report, released in 1979, and laid the foundations of modern, international research ethics rules and laws. The 'Belmont Principles', which define the duties of scientists to human subjects, include respect for persons, beneficence, justice (involving vulnerable subjects and populations), fidelity (involving fairness and equality), non-maleficence, and veracity. These principles, which owe their genesis to 2,000 years of philosophical study of ethical theory, have come to be applied through a variety of professional, national, and international rules, standards, codes, and laws.

Given the lapses by scientists, governments, and corporations in permitting harms that came about not due to human subjects research, but through the development of technology and marketing of products outside of medicine, might we consider applying the Belmont Principles to *scientific research in general*? I have argued as much in the journal *Science and Engineering Ethics*.[7] Specifically, if the Belmont Principles embody ethical duties owed to direct human subjects of research, shouldn't they also be applicable to *humanity* as a whole, or at least to those who are directly affected by all research through the development and introduction of products into the stream of commerce? For instance, workers exposed to hazardous chemicals during industrial processes, and who were not properly informed about the nature and consequences of those exposures, were not able to give fully *informed consent* to their exposure. Consumers as well, who purchased Pintos without foreknowledge about their potential for exploding in a rear-end collision, or that Ford could improve each car (or a customer could, with knowledge) with an inexpensive after-market upgrade, were treated unfairly, violating the manufacturer's duty of fidelity. Why should the Belmont Principles only apply to duties owed by researchers to human subjects in studies, and not to all scientists and developers (including those who move research into practical engineering, and thus into the stream of commerce) performing research that may have direct effects on consumers? As a modern example of the extent of damage that can be done by both scientists and non-scientists making decisions based upon the current state of scientific understanding, consider the BP oil spill in the Gulf of Mexico. Numerous technical fixes could have helped prevent the blowout that resulted in the largest historical release of oil into the oceans that killed eleven of BP's own workers, and that now threatens a generation of workers and residents in the Gulf region. Careful attention to the principles enunciated in the Belmont Report, including the need to treat people with dignity, and to avoid unnecessary risks, might have helped put the conflicting desire for profits in its rightful place.

Nanotechnology, like all technologies, poses the potential for both significant improvements in our well-being and lifestyles, as well as the potential for harms. These harms may be environmental, as discussed above, or they may be direct harms to those working with the materials involved, and consumers who willingly purchase nanotech-based goods. Expanding the moral horizon of the Belmont Principles to humanity as a whole, which is, after all, subject to the introduction of new technologies, even though we are not all subjects of studies, should help us to avoid some of the ethical lapses of the past. Whether nanotech faces the same future as medical research depends largely upon those who are working on the basic science and those who will bring its products to market. While a fair amount of the institutional machinery now involved in bioethics involves self-policing and peer-review, given the lapses of the past, governmentally enacted rules, regulations, and laws now back up many of those institutions. Failures of research ethics in the modern era can bring legal consequences, or at least, result in significant institutional punishments, fines, and withholding of licenses, as well as personal and professional liability.

Self-policing of behavior is preferable in modern liberal polities and economies, both because doing the right thing initially out of proper motivations is morally preferable to either skirting duties, or acting with bad intentions, but also because it is more efficient. The lesser the bureaucracy that is necessary, the lower the costs of the technology or industry, and institutional rules and regulations add to the bureaucracy. Modern human subjects research, because of Tuskegee and other lapses, is very expensive, costing researchers and institutions time and money to oversee the ethics of their ongoing studies. This is inefficient, but necessary, given that we have seen the dire effects of even a small number of un-policed professionals acting unethically. Tuskegee could have been avoided had researchers acted with greater regard for their ethical duties, setting aside their scientific enthusiasm momentarily to reflect on duties owed outside of science, to the subjects of their research.

It is incumbent upon both scientists and those seeking to create nanotech-based products that either affect the environment or enter the stream of commerce and directly affect users and those working with those products, that the mistakes discussed above be avoided. The ethical duties embodied in the Belmont Principles and similar, international codes are owed despite the regulations that came to be deemed as necessary. To avoid similar, top-down rules, regulations, and laws, and the bureaucratic inefficiencies that inevitably follow, choosing at the early stages of the development of nanotech to abide by ethical duties may help obviate the need later to enforce good behavior by institutional means. Even absent a desire to do the right thing for its own sake, enlightened self-interest should provide sufficient incentive to avoid the mistakes of other sciences and industries. As discussed above, public over-reaction by applying the precautionary principle to GMOs in Europe was likely precipitated in part by apparent lapses in other industries.

Proper respect for the ethical duties noted above requires special consideration of the characteristics of nanotech products, and why they must be carefully studied before people are exposed to them. The factors that make nanotech so interesting and useful, namely the size of the materials and machinery involved, make nanotech a special concern regarding human exposure. Specifically, small things are more 'reactive', and pose the possibility of harms that larger materials do not. Because of their high surface area, nanoscale materials and objects may be inspired through airways, become more deeply lodged in lungs and other tissues, and even permeate the skin, all offering means to harm people that many other products do not. Various organizations, professional groups, and governments have recognized the special nature of nanotechnology in regard to human health, and have promulgated various recommendations and codes of behavior to guide the fledgling industry. The Foresight Institute, which owes its genesis to Eric Drexler, published their 'Foresight Guidelines for Responsible Nanotechnology Development' in 2006.[8] In 2008, the Commission of the European Communities published their report, 'A code of conduct for responsible nanosciences and nanotechnologies research'.[9] Each of these reports expresses interest and awareness of the potentially unique risks posed by nanotechnology, as well as the desire to ensure that the benefits are realized through responsible development.

It remains to be seen whether nanotechnology will suffer lapses like those of medical science, and whether more regulation or other institutional responses will eventually be necessary to protect people from its potential harms. While nanotech poses unique harms, as noted above, these features have been present (and proven harmful in some circumstances) with other technologies and products. Dioxins are molecules, and cigarette smoke (and other pollutants) are nanoscale. So while we are familiar with the risks in general, each new nanotechnology product will pose unknown risks that must be judged carefully. And while we can agree that the community of well-intentioned researchers and developers of marketable technologies will do well to be guided by ethical principles, it is concerns about bad actors, whose intentions are already unethical, and who *wish* to cause harm, that leads us to consider whether rules, regulations, and laws should govern the dissemination of technologies (like some nanowares) that can be put to evil purposes, by individuals or groups who seek to harm others. And if so, how?

Security: Can and Should We Prevent Evil Uses of Nanotechnology?

Even the most benign technologies are eventually adapted in order to cause harm. Commonly available fertilizers meant to increase crop yields killed many scores of innocent people when Timothy McVeigh used them to bomb a US federal building in Oklahoma City in 1995. Household implements of every conceivable kind have been used to hurt, maim, or murder people for as long as man has been making tools. Certain technologies, however, have been developed recently that are considered to be so inherently dangerous that significant regulatory apparatuses have been developed to contain them.

Until the twentieth century, guns and gunpowder were generally available to anyone who could afford them, but larger, deadlier weapons such as canons remained too expensive for those of ordinary means. While many technologies had been developed specifically for killing and warfare, techniques that incorporated the use of deadly chemicals into ordinary arms initiated the first attempts to curtail the use of certain technologies relating to warfare. As early as 1675, the Strasbourg Agreement between France and Germany banned the use of poison-tipped bullets. In 1874, the Brussels Convention regulated the use of chemical weapons. Although three Hague treaties were signed before the start of the First World War, chemical gas weapons were used in that war, and have been occasionally used even since the signing of the Treaty of Versailles, which in Article 171 stated that 'the use of asphyxiating, poisonous or other gases and all analogous liquids, materials or devices being prohibited, their manufacture and importation are strictly forbidden in Germany ... the same applies to materials specially intended for the manufacture, storage and use of the said products or devices'.[10]

After the Second World War, control of chemical weapons was subsumed into general, international regulations concerning weapons of mass destruction, which included the newest, deadliest technology: nuclear weapons. Fear about the spread of nuclear technology inspired the United States to make the technology itself classified, and to forbid patents on it as well. One of the purposes of patents is to enable others to practice and improve the art disclosed in its claims once the patent expires and the invention moves into the public domain. Fear about the proliferation of nuclear weapons, and the use of nuclear technology by other states to create their own weapons, encouraged secrecy and restraint of the technology itself, except by those within the US government. But knowledge about the underlying science was already well-known, and other governments soon duplicated the United States' success in building both fission and fusion bombs. The genie was out of the bottle. Yet even after the spread and duplication of nuclear technology by other states, the United States, and most of the other nuclear states, attempted to regulate proliferation of both the knowledge and the production of nuclear materials, and eventually entered various treaties among themselves to further limit the proliferation of nuclear technology outside of the select group of first-comers.

Nuclear arms control and anti-proliferation treaties have created international monitoring and enforcement mechanisms to track the flow of fissile materials, and to outlaw attempts to build or otherwise possess nuclear weapons by other states. Similar treaties continue to monitor pathogens capable of use in biological warfare, and also the stockpiling of dangerous chemical agents. Since the Oklahoma City bombing, even quantities of fertilizers capable of being used for explosives are now tracked, and limits on who can purchase them are imposed by various laws and regulations. While uranium is found in the environment, purifying it for use in a weapon can hardly be done without attracting the attention of those agencies and organizations tasked with tracking the attempts to build nuclear weapons. It is expensive, complicated, and takes attaining a certain level of technological advancement, and the possession of specialized equipment, to create fissile materials for bombs. Manufacturing (without being noticed) sufficient quantities of chemical weapons for use as bombs or for terror also is difficult. Creating weaponized bio-warfare agents is easier still, and harder to track. Witness, for instance, the successful anthrax attacks in the United States in 2001 and 2002.

Nanotechnology and synthetic biology pose potentially catastrophic possibilities for rogue states and terrorists to attain weapons of mass destruction without drawing attention, and cheaply. Because of their potentials for essentially 'garage' or 'basement'-made mayhem, those who are involved at the early stages of these sciences are also trying to develop voluntary ways to track the use of at least precursors to deadly products. Nanotech terrorism is a long way off, though arguably the weaponized anthrax may have been purposely coated with silicon particles.[11] In the meantime, synthetic biology is receiving the attention of those in various militaries and security agencies, and by the researchers themselves who understand the full range of possibilities.

Security and Synthetic Biology: Precursor to Nanotech?

Synthetic biology is essentially engineering at the nanoscale using biological systems. As we have noted above when discussing the feasibility of true molecular nanotechnology, proponents have argued that since biological systems construct highly accurate nanoscale structures, there's no reason to think humans could not engineer similar systems. Systems biology, or synthetic biology, essentially extends an engineering approach to biological systems, attempting to create basic building blocks by altering genetic code to construct materials and even nanoscale machinery. One of the essential mechanisms for synthetic biology is the identification of useful snippets of genetic codes, and other useful biological materials, so that they can be combined in new ways. One can now order custom strings of DNA from companies that sell them to researchers, or if one can afford the equipment, create the sequences oneself. In 2002, researchers at SUNY Stony Brook created a synthetic polio virus, pathologically identical to a naturally occurring one (but with markers to denote its artificial manufacture), by using mail-ordered sequences.[12] The potential for mischief as synthetic biology matures is clear from the polio example. If polio could be constructed in the lab, then so could smallpox, or even worse, hitherto unknown biologically based weapons of mass destruction.

Synthetic biology is being touted as a quick and easy path to some of the promises inherent in nanotechnology, piggybacking off the success of nature in designing nanoscale processes and products, and speeding our means to achieve nanoscale constructions of our own. The tools to make it possible are also becoming cheaper, and more available to 'garage' and 'basement' amateur (or professional) synthetic biologists. With spreading and growing knowledge of the fundamentals of biological processes, combined with falling costs of equipment and greater availability, states are growing nervous about potential uses by terrorists, and researchers understand that such an incident, should one ever happen, will bring this fledgling science to a screaming halt. Self-policing the industry by a variety of mechanisms has become a widely accepted necessity, even if there are questions about its practical efficacy.

In 2002, researchers in the field met for their second international conference, 'Synthetic Biology 2.0', and discussed in some depth issues relating to safety and security. In 2006, out of that meeting and subsequent meetings and colloquia (as well as public input) a white paper entitled 'From Understanding to Action: Community-based Options for Improving Safety and Security in Synthetic Biology' was published.[13] The document heavily stresses the duties of researchers and commercial suppliers in the field to be aware, and to self-police. Numerous other efforts by NGOs, governmental bureaucracies, commissions, and law enforcement agencies have also begun to examine the security implications of the spread of knowledge and means to conduct synthetic biology. Several national and international consortia in Europe have launched inquiries and studies into the ethics and practical concerns of regulating synthetic biology, with a special emphasis on security concerns. In 2007, a white paper was published by Synbiosafe, a project involving the University of Bath, the University of Bradford, and the Organization for International Dialogue and Conflict Management, all partners in the Synbiosafe project.[14] In 2009, the European Group on Ethics published its report on ethical and practical issues in synthetic biology, noting certain security issues as well.[15] The Synth-ethics project funded by the EU, and on which I have been an investigator, published its first report in late 2010.[16] The trend, begun with the meeting of Synthetic Biology 2.0, is to focus on voluntary notification and enforcement mechanisms, as well as individual researcher responsibility. This model is proposed by a joint report of the J. Craig Venter Institute, MIT, and the Center for Strategic and International Studies (CSIS). The report, published in 2007, is entitled 'Synthetic Biology: Options for Governance'[17] and it explores a number of policy options. Although it presents no recommendations *per se*, the weight of the projected impacts of the various options presented leans heavily in favor of voluntary professional oversight, education, and openness, in order to encourage innovation without significant top-down control mechanisms, and to help prevent intentional misuse of the technology. The alternative to this trend is much more closed, tightly regulated research, including regulation affecting knowledge dissemination, and top-down, government oversight of labs, precursor materials, and researchers. Of course, this sort of restricted environment is generally anathema to a liberal democracy, to the smooth conduct of research in a rapidly evolving field, and it is doubtful that it would accomplish the overall goal of improving security.

As we have noted, the knowledge and materials necessary for synthetic biology are already generally available, and growing cheaper every day. Unlike the tools and materials used in nuclear weapons, chemical weapons, and even some weaponized biological agents, there is really very little that can be done to effectively police the pursuit of synthetic biology, which is now literally possible to conduct in garages and basements. For now, this is not the case with the tools and knowledge necessary for pursuing true molecular nanotechnology. So far, the cost of things like powerful electron microscopes is prohibitive for garage tinkerers, and the various grassroots efforts at creating fabricators are not approaching anything like the technical detail involved in molecular nanotechnology. But if the trajectories of both the top-down and bottom-up approaches to nanowares continue to merge, then nanotechnology will have to take issues of security as well as safety seriously, as researchers in its sister-science synthetic biology are beginning to do. Perhaps more so than any conceivable industrial failure, either intentional or accidental, the use of nanotechnology by bad actors, intent on causing significant harm, will undermine public confidence in the technology, and bring to bear a measure of government regulation and oversight that could choke the science and technology, and hinder its progress and benefits.

What can we learn from regulatory efforts in other, dangerous technologies in the past, and how can we best pursue nanotechnology's benefits while avoiding the environmental, safety, and security concerns expressed above? And what other regulatory and governance issues can we best serve now, in the nascent stages of nanotech science?

The Path of Openness

Consider what might have happened had nuclear technology been kept open, and the knowledge and means of producing nuclear weapons, as well as nuclear's peaceful uses, not been regulated so heavily. Would the world have been less safe? At one point during the Cold War, when the Soviet Union and the United States had helped regulate an international climate in which those two states held a virtual monopoly on nuclear weapons, each side had enough warheads to destroy the earth. A nuclear exchange would have eliminated most life on the planet. How safe were we? During the Cuban Missile Crisis, we were closer to nuclear war than at any other point so far in history. A diplomatic failure could have spelled the end of civilization. War was only narrowly averted. The balance of terror maintained by the policy of mutually assured destruction (MAD) may have helped avert nuclear war, or it may have been just lucky that given our capabilities, we conscientiously avoided use of our nuclear weapons due to some other inhibition. In the post Cold War era, some truths have emerged that test the value of the MAD policy, and that suggest that deadly technologies will not be likely to be used even by rogue states or terrorists, with some exceptions.

Although nuclear technologies continue to proliferate, and now states like Pakistan, India, Iran, Israel, and North Korea all possess, or likely possess, either the technology to produce nuclear devices or the devices themselves, they have not yet been used. International non-proliferation treaties and policies of containment have generally failed. These agreements actually serve as bargaining chips, rather than deterrents. As a society attains the level of technological capability to produce nuclear weapons, it makes more sense to do so secretly, to the degree one can, and then to use this capability once achieved to bargain for something. International pressure to limit proliferation of nuclear weapons creates a climate for blackmail. States that skirt these agreements and develop their own nuclear capabilities can then taunt the world community with their technological achievement, flaunt their violation of treaties, and use their new membership in the nuclear club as leverage to secure aid, cooperation in some other dispute, trade deals, or other demands. Since the end of the Cold War, by which the two major nuclear superpowers essentially stalemated each other, the growth in the number of nuclear states has been steady, and the threat of MAD has failed. Instead, international efforts to curtail the spread of nuclear technology seem to have achieved the opposite. And yet, are we any less safe?

The post Cold War era offers a glimpse of what the world might have been like had we never regulated nuclear technologies. If everyone has weapons of mass destruction, is the threat of nuclear conflagration any greater than if only two mortal enemies possess them? The risks of using nuclear weapons for the state that chooses to use them seem to increase in a world in which everyone may retaliate with nukes, as opposed to a world where only the few states that have monopolized the technology might retaliate, and likely without nukes. In the Cold War world, if the USSR or the United States used nuclear weapons on a small state that had developed and used a nuclear weapon on its non-nuclear neighbor, the chances of a US/USSR nuclear exchange would have increased, and either of the two superpowers would have looked like a bully. In a post Cold War world, in which (presumably) anyone might develop and possess nuclear technology, the risks to any state that chooses to use nukes increase dramatically, because retaliation could be immediate, and pose less diplomatic shortcomings to non-superpower states that choose to use them, given that such a use would be self-directed and legitimately defensive.

Had the nuclear world after the Second World War been multi-polar, instead of bi-polar, and had no caps been imposed upon the research and development of nuclear technology, an international stalemate would have likely prevented the wartime use of nukes. We are arguably less safe living with the knowledge that anyone might develop destructive technologies in secret, then blackmail us later, than if we simply assumed that anyone might develop and possess weapons of mass destruction legally if they possess the capability. The latter climate encourages multi-polar diplomatic agreements to curtail the *use* of these weapons, rather than complex institutional measures and threats of force to prevent their *development*. The former, as we have seen, incentivizes secretive research and development, and then dramatic public blackmail. There seems more to gain, for safety and security, from openness than through tight regulation and curtailment of knowledge and technology.

Astonishingly, we have failed to destroy ourselves as a species, despite the means to do so held for the past sixty years. Of course, it's still possible that we will do so, just because of the still vast proliferation of nuclear weapons, primarily still in the hands of the United States and Russia. Every day we continue to move forward, and lately, to continue to reduce the numbers of nuclear weapons, the likelihood of global nuclear catastrophe falls. This doesn't mean that someone won't someday use a nuclear device in war (as the United States did twice) or for terror. But cheaper, easier means for terror exist, as the events of 11 September 2001 graphically demonstrate. Doomsday scenarios aside, more banal and significantly destructive means of killing people remain widely available, and no amount of regulation can rein them in.

Could it be that we can be trusted with dangerous knowledge? Might even the most evil of men be prevented by external factors, or by fear and trepidation, from engaging in deliberate acts of large-scale destruction? We can hope, though history shows that outliers emerge now and then who will stop at nothing to kill, or to commit genocides, or launch dreadful wars. And while scientists and innovators must be cognizant of the possibility that such people will use their technologies for harm, this cannot serve as an argument not to pursue potentially deadly knowledge. Attempts to build nuclear weapons were underway inside Nazi Germany and outside, the knowledge of Nazi efforts to do so, and Albert Einstein's knowledge of the capability of the Nazis technically to succeed, arguably helped enable the Allied forces to prevail. No chance of prevailing could have emerged from attempts to squelch the knowledge itself.

It is the nature of information and knowledge to spread, despite attempts to curtail it. Attempting to curtail the spread of seemingly dangerous knowledge only encourages those who wish to have that knowledge at all costs to do harm to acquire it, and to operate underground, secretively, and beyond the view of those who might be able to prevent that knowledge's evil uses. Consider the drug trade. A dangerous underground system of manufacture and distribution exists, by which thousands of people are killed each year in wars among rival gangs, and the products of which are unregulated, impure, tainted with the blood of innocents, and uncontrollable. Market demand has continued unabated, and even been exacerbated, despite and perhaps due to regulations. As users are criminalized, the ability to intervene and treat addictions is diminished, and the cycle of illegal, underground use, manufacture, and distribution is that much harder to track. If people want something, they will find a way to make it, or entrepreneurs will emerge who will satisfy a market demand.

Attempts to curtail knowledge about nanotechnology, or to regulate the availability of the machinery and equipment needed to realize its full potential, will ensure that a black market emerges. It will become that much harder to track the development of potentially harmful products and uses, and overall safety and security will be diminished. We should instead encourage openness. We should view the Cold War, and even more so, the post Cold War international stalemate as evidence of success. Given the means, the better angels of our natures can be trusted to prevent intentional catastrophe. The distant possibility of grey goo should be kept in mind, and the current and near-future dangers of synthetic biology ought to motivate us, but only to educate those who are involved in these sciences about their duties, and to encourage the free spread of knowledge as a *preventive* measure. The more we know about the possibilities, and the better capable we are of evaluating risks, the more likely the community of researchers will be motivated and able to prevent technology's harmful uses.

Openness also directs us to more proactive measures, besides guiding our protective actions. Considerations of *justice* should encourage efforts to repudiate regulatory measures aimed at curtailing the free flow of information wherever it threatens the positive potential of the technology. As we have discussed, the *modus operandi* of liberal democracies is to increase political participation, encourage freedom, and open markets. Yet powerful regulatory forces currently work against these goals out of an expressed motivation to encourage the progress of 'science and the useful arts'. IP is now taken for granted as a right, although its history suggests that the rights established by IP are relatively recent, wholly positive, and not founded upon the sort of principles that have grounded other human rights. If we are interested in promoting the growth and full potential of nanotechnology, and investigating the role of all regulations in this effort, then we must also focus on the role and impact of IP on innovation in general, and nanotechnology in particular. We cannot take for granted that it always achieves its stated ends, nor that we must accept without modification its current forms.

IP frameworks, initially established to provide an incentive for authors to create and inventors to innovate, are state-sanctioned monopolies that may hinder innovation. They are, essentially, regulations that curtail the use of knowledge by granting to a first-comer an exclusive monopoly, guaranteed by the force of law. In the next chapter, we will inquire further into the nature of intellectual monopolies, the theoretical underpinnings of IP law, and what specifically these laws mean for the development of nanotechnology in light of concerns for justice.

CHAPTER FIVE

Things in Themselves: Redefining Intellectual Property in the Nano-age[1]

We have been confronted in the last 100 years with a rapidly changing technological environment. This environment has brought with it great wealth and prosperity, and there is no doubt that emerging concepts and applications of intellectual property (IP) have aided to some degree our scientific, technological, and economic expansion. Copyrights and patents offer monopolistic rights to authors and inventors over their creations, ensuring profits for fixed terms, and providing fortune as well as fame for the most successful inventors and authors. Monopoly rights are strong incentives to create both utilitarian and aesthetic works.

Recently, emerging technologies have challenged traditional IP theory and practice. Consider the rise of computerized phenomena, and the proliferation of software and instantaneous communications through the internet. Software has been considered a kind of hybrid, treated both as a patentable invention and as copyrightable expressions. In my first book, *The Ontology of Cyberspace*, I considered how software has revealed that our distinctions between copyrightable and patentable objects are artificial and illogical. I argued that all intentionally produced, man-made objects are expressions, and that computerization has merely revealed how one expression is very much like another. More recently, I have considered the question of patenting unmodified genes. In the course of that discussion, I further criticized the application of existing IP to genomics and genetics, at least where patents have been issued for unmodified genes or life-forms. In all of my work, I have sought to uncover the ontologies (descriptions of the objects and relations involved) of the underlying objects. I have done this believing that once we reveal the nature of things (like expressions, machines, software, and genes, as well as relations and social objects like property, ownership, and intentions) we could then sensibly sort out logical errors and inefficiencies in public policies.

We are now on the cusp of a new engineering breakthrough that will once again challenge our relation to our technological world, and likely pose new challenges to the application of traditional IP. Nanotechnology involves the construction of materials and objects at the 'nano-scale', beginning at the molecular level. Theoretically, this will mean cheap and abundant objects of any size and shape, more or less instantly created, with little to no waste, and constructed anywhere and at any time. The science-fiction notion of simply dialing up an object and having it constructed on the spot, molecule by molecule, may well be decades away or further, but it is the ultimate goal of many who pursue nanotechnology research. Even so, we will begin facing unique challenges about the nature of IP as the dividing lines between ideas and expressions further blur, and matter itself becomes programmable.

IP has become a major force economically and culturally, impelling in part our hyperbolic economic and technological growth in the past century, and influencing both cultures and economies worldwide. Historically, the emergence of IP as a class of objects is relatively new. This class of objects, which includes things that are patentable, copyrightable, and trademarkable, has evolved since its inception both significantly and rapidly. Let's look briefly at the evolution, form, and purposes of IP law, and then consider how and to what extent new material technologies will finally destroy current notions of IP.

The Emergence of Intellectual Property

IP is often mistakenly thought to protect ideas in much the same way that ordinary property law protects things. In fact, there never has been any legally recognized protection for ideas themselves, but only for various forms of expressions. Beginning around the time of the Renaissance, in Britain and Italy, the first legally sanctioned monopolies over inventions and works of authorship were created. Some of these monopolies took the forms of 'letters patent' entitling inventors to a limited monopoly for the sales of their inventions. Others were in the form of copyrights, entitling first publishers to monopolies over printed works, and much later, moving that monopoly right to authors and artists. Over time, IP protection became formalized throughout the western world, and embedded, among other places, in the grant of powers to Congress in the United States under Article 1 Section 8 of the US Constitution.[2]

This emergence was not without its periods of reticence, and particularly during the nineteenth century, major economies in western Europe drew back from their recent embrace of IP, and went through periods where they completely rejected the notion of IP. While France and the United Kingdom had been among the first and most eager adopters of IP regimes, following the example set in the United States following the revolution, the United Kingdom had a long-lasting intellectual debate about the wisdom and necessity of IP in general. France, among the European nations, had most fully succumbed to their particular arguments about the *moral* rights of authors and inventors, establishing their patent and copyright system early after their revolution, and never backing down. But in the United Kingdom, propelled by modern economic theory, and given prominent voice by the journal *The Economist* and others, anti-patent sentiment grew. The anti-patent movement, which will be more thoroughly detailed later, culminated in lengthy periods in the mid- to late nineteenth century when the Swiss, Germans, Dutch, and British did away with IP laws entirely, even as their economies apparently continued to grow.[3]

The theoretical underpinnings of IP law were thoroughly debated during this anti-patent period, and four major arguments promoted the use of patents: (a) there is a natural right to the products of one's inventiveness, for example, ideas; (b) society should reward inventors for their service by grant of some limited monopoly; (c) economic and technological progress is good and will not be encouraged without some grant of a limited monopoly; (d) government-granted IP rights require disclosure, which benefits society and which would not exist without IP laws. Inherent in these arguments, and generally adopted by proponents and opponents of IP laws alike, is the understanding that the *nature* of the subject matter of IP *differs in kind* from other forms of property. Ideas tend to spread without control, and it is difficult or impossible to naturally defend them absent secret-keeping. Even secret-keeping does not prevent independent discovery. Nevertheless, in ordinary marketplaces, it is possible to see and track the movement of goods, and generally also the use of various processes. If governments create IP systems, patent holders can seek redress if they see their techniques or technologies reproduced without license, just as authors can note substantial similarities between their works and the works of copiers, and sue them in court for violating their IP rights.

The creation of government-sponsored monopolies over types of expressions is intended to balance two competing goals: encouraging innovation and moving knowledge into the public domain. Historically, secret-keeping was one of the first means by which inventors and, to some extent, authors profited from their creations. Trade secrets are still an effective means of preventing others from profiting for one's unique inventions, but only where those inventions are somehow impenetrable to 'reverse-engineering'. As technology and science make the process of 'reverse-engineering' cheaper, easier, and more widely available, secret-keeping has diminished as an effective means of protecting the fruits of one's inventions. In order to ensure that authors and inventors would continue to contribute to society, governments have established IP regimes to ensure guaranteed profits for the fruits of authorship and invention for certain time periods. Currently in the United States and elsewhere, authors are granted a monopoly while they are alive, and their estates still profit for seventy years after the author dies. Inventors only get a twenty-year monopoly following the filing of a patent.[4]

Among the justifications for granting such exclusive monopolies is the assumption that without guaranteed returns on investment (of time or money) neither inventors nor authors would take the risk of innovating or creating. Behind this pragmatic goal are some notable theoretical underpinnings, some of which have been challenged of late by new technologies. A central theory of IP is that ideas and expressions are distinct from one another, and no monopolies should ever limit the dissemination of ideas. The idea/expression dichotomy, as it is called, recognizes that the ultimate goal of IP law is to benefit the public, even while benefiting inventors and authors for their unique contributions to society. The idea/ expression dichotomy is grounded in sound metaphysics. Not only are there practical distinctions among things and ideas, but there are also uncontroversial ontological distinctions between, for instance, a chair and the *idea* of a chair. If you don't believe me, just try sitting on the *idea* of a chair.

Ideas versus Expressions

The idea/expression dichotomy may also be behind some prominent, recent confusion in the treatment of emerging technologies under IP law. As we have touched upon briefly above, there have been historical prohibitions against granting *too much* protection to certain objects, notably, 'laws of nature' and ideas. Laws of nature and 'abstract ideas' cannot be monopolized for two reasons: (a) they are not the products of human authorship and (b) granting monopolies over them curtails useful arts derived from them, which contravenes the general purposes of IP law.[5] Ideas cannot be monopolized for fully pragmatic reasons as well. If ideas were amenable to IP protection it would pose a similar problem of stifling innovation, as well as one of enforcement. If a monopoly existed over an idea, rather than only its expression, how could one prevent others from expressing that idea in all potential media, or from just being uttered? Protecting bare ideas would be unenforceable. Instead, given that IP protection only extends to expressions in *fixed* media, rather than the bare ideas, enforcement is limited only to instances where an unauthorized expression appears in the market. But there are certain expressions that appear at first to be much like ideas, if not ideas themselves. It has been objects like these that have resulted most recently in problematic treatment of new technologies where authors, inventors, and those who apply the laws of IP have treated some objects as erroneously protectable, and others as unprotectable. Historically, these distinctions were not so problematic, until modern technologies blurred these lines.[6]

As patent and copyright law developed, it was clear that machines and processes were patentable, whereas works of authorship were copyrightable. Each was an expression of an idea in some medium, the primary roles of each differed socially. Works of authorship were not primarily 'utilitarian' whereas an invention was. Books seemed very different from the printing presses that made them, and even more clearly distinct from the ideas that conceived each. Ideas, and expressions, either as machines or as works of authorship, were most clearly distinct when media were less complicated. As media have evolved, those distinctions have blurred. A painting expresses an aesthetic idea in a traditionally artistic medium, but what about computer art? When digitized, a painting resides in computer chips, as the logic of electronic gates as part of a machine. In theory, the same painting could reside in mechanical computers as well, as parts and processes of obvious machines. All of which requires us to ask the question: In what way does the medium matter when an idea is expressed? We could logically conceive of a machine whose sole purpose is to store and express a particular painting, let's say, the Mona Lisa, just as an optical disk can store and express the Mona Lisa. How does the medium alter the nature of the painting, if at all? What makes paint different from computer chips, mechanical or electronic logic gates, or bits on an optical disk?

As media have become more flexible, and the ability to express the same ideas in various disparate media has become possible, laypersons, lawyers, and jurists have had difficulties distinguishing among ideas and expressions.

The blurring has actually been of the historical distinctions between machines and other, 'non-utilitarian' expressions, but has bled over into the idea/expression dichotomy. Transmissions over airwaves or internet, at the speed of light, *seem* to many to be like ideas themselves. This is but a convenient delusion for many, and for others it is an actual ontological error. I have detailed the delusion of Michael Heim, in his *Metaphysics of Virtual Reality*, in which he continually conflates cyberspatial entities and ideas themselves.[7] 'Virtual' artifacts are not, however, ideas. They *seem* like ideas only in their evanescence. Turn off the computer and the expression stops. Turn off the monitor and the painting disappears. Wipe the memory clean and the virtual artifact cannot reappear.

Conflating ideas and expressions in new media serves a number of ideological interests as well, interests with which I am not wholly unsympathetic. But ideology is not a sound basis on which to make decisions about ontology. Ideas differ from their expressions in one very important manner. Ideas can be held in the mind of one individual without becoming expressed. Once an idea leaves a mind, it is either expressed or forgotten. If forgotten, then the idea itself no longer exists as a thought. If expressed, then other minds may come to possess the same or similar idea, depending on the success of the expression. There is no sense in which any object that exists in a computerized medium is an idea because, so far, artifacts are not yet able to conceive things. All computerized artifacts are (also, *so far*) the intentional results of humans. This does not rule out the eventual possibility that minds will come to exist in artificial media, but until then we have no right to conflate ideas with virtual artifacts. This conflation, while popular, must be seen for what it is: ideologically motivated.

Marshall McLuhan's oft-misquoted 'the medium is the message' (it was originally the 'massage') adopted by proponents of the 'new' media and a certain anti-intellectual bent are partially responsible for spreading a sort of new-age metaphysics about cyberspace.[8] This metaphysics, adopted without much scrutiny by Michael Heim, and spread among those who would leave cyberspace an ungovernable medium, immune to standard IP regime, assumes that virtual objects *are* idea-like, if not actual ideas. Heim goes as far as to call virtual objects 'pure thought' or Platonic ideals. But ideas are unique. So far, they can only subsist as thoughts in the medium of a mind, which for now is a biological brain. Of course, mind need not necessarily be limited to biological brains, but until there is better evidence of true, thinking artificial intelligence, capable of forming new ideas and expressing them, we cannot logically confuse ideas and expressions in artificial media. Nothing justifies it.[9]

Our standard idea/expression dichotomy can hold for now. Ideas differ from expressions, which are always the instantiation or depiction of an idea by way of artifact or intentional performance. Ideas cannot be pragmatically protected by IP laws because there is simply no way to restrict an idea to a thought in just one mind. Once it has become expressed, other minds may then hold the idea. Moreover, nothing prevents independent 'discovery' of an idea by any other mind. So IP law protects only expressions. It gives a limited, exclusionary right to the holder of the IP, preventing others from expressing the protected IP until a certain time. This has worked historically, largely because of the enforceability of IP regimes over old media. Pages were harder to copy than files, and widgets more difficult to reproduce than digits. But the new media are attractive because of their portability, speed, flexibility, and ease of use. Now, IP laws are violated, and unauthorized copying abounds in cyberspace. This raises now some vital questions, which I have set forth before, but which deserve exploring as the new media continue to alter the means of production. Can standard IP regimes adapt to the new media in sensible ways that encourage innovation, but do not make criminals of ordinary people?

The answer in cyberspace has been yes, but it has occurred at the grassroots. Governments have not even sought to change the fundamental modes of IP protection, choosing instead only to increase the length of terms of protection, and stiffen the penalties for violations. Rather, innovators, sensing the losing battle involved in strong IP, which is often a tool of the already wealthy, and seeing it as a hindrance to consumer acceptance and robust product development, have chosen another path. Alternatives to standard, government-sponsored monopolies have emerged, and have been successful, employing private contract and licenses. Creative-commons, open source, shareware, GNU/GPL, and other innovative licenses have evolved to cope with the inherent strengths and weaknesses of new media, to encourage their economic exploitation, and to not treat users as potential enemies.[10]

The new media have revealed flaws in our old distinctions between patentable and copyrightable objects. The distinction is rooted in pragmatic concerns, and a certain preference, it seems, for utilitarian objects (inventions) over aesthetics. But the distinction ultimately is among objects without any ontological difference. Inventions and works of authorship are all expressions, the uses of which differ socially. Works of authorship are expressions (the extensions of ideas outside minds) with primarily aesthetic uses, and inventions are expressions with other types of utility. New modes of IP protection should be agnostic, ultimately, about the medium, recognizing that what matters in order to encourage innovation, and to retain consumer good will, is access to good, flexible, open goods at reasonable prices rather than governmentally supported monopolies. Artifacts can be conceived, designed, produced, and delivered utilizing the new media in cheaper, often distributed modes. Individual innovators can reap rewards from consumers not just in the margins they realize with each individual copy delivered, but also through an ongoing, open design phase so that consumers become co-producers, sharing in the profits by better products, and at reduced prices. New media have not only revealed flaws in the original conception of IP, but also encouraged the creation of new modes of distribution, and grassroots responses to IP laws that seemed to hinder entrepreneurial activities.

But is this new wave of innovation, which has seen success with products like Open Office, Mozilla Firefox, Apache server software, and others, limited to the domain of bits? What happens when atoms become like software, programmable, instantly distributable, and abundant? This is the future envisioned after all by a growing number of those working to fully realize nanowares. Will IP laws as they have existed so far suffice for such a future?

Atoms for Bits: Pragmatic and Theoretical Challenges

For the rest of this chapter and the next, I will concentrate on the applicability of current theories of IP to nanowares. Adapting or altering current IP schemes to fit emerging modes of manufacturing and distribution requires us first to have a coherent theory of artifacts and ideas, and then a useful mode of organizing our behaviors so as to promote the goals of IP laws. This first part is largely accomplished, though not necessarily explicitly. Implicit in the law of IP is a sound ontology of artifacts, expressions, and ideas. Although the current regime complicates matters somewhat by artificially dividing the domain of expressions into 'utilitarian' and 'aesthetic', as mentioned above, this distinction only truly became problematic when the new media emerged. Nevertheless, underlying IP law has always been a recognition of the critical ontological distinction between ideas and expressions. Expressions are ontologically dependent upon ideas, but not vice versa. Ideas can be held and never expressed. Ideas are ontologically primary and predicate to an expression. Expressions of ideas are intentional manifestations of those ideas beyond a mind.

The chain of ontological dependence and priority begins with minds, which conceive of ideas. Ideas, so far, subsist only in minds as thoughts. Thoughts are material components of mind, and while they are evanescent, they exist in certain locations at certain times. It is their evanescence and their privacy, as well as a potential preference for freedom of thought, that prevent us from allowing monopolies over ideas. Moreover, the similarity of bare, unexpressed ideas to other ideas in other minds cannot be determined without some expression, so not only is enforcement of trespass on ideas difficult, but isolating those trespasses without some further act seems practically impossible. As touched upon above, nothing about the nature of minds or ideas requires that ideas be located *only* in human brains; it's just that so far we only have evidence for their existence in human brains. It is possible that other animals, or eventually machines, could conceive ideas, and even express them. Meanwhile, the IP regimes under which we work only apply to expressions that originate in human minds, and only human authors or inventors get the benefits of protection. It is only human ideas in human minds that we have evidence of through every single human artifact that has ever been created.

All expressions are *products* of human intention. Some expressions are artifactual, others are less clearly so. A dance is an expression, as is an utterance. But these sorts of expressions, because they are fleeting, are ordinarily never accorded any sort of monopolistic protection in IP regimes. Representations of these sorts of expressions in some medium which lasts over time are often afforded IP protection. There are both theoretical and pragmatic justifications for this. Non-fixed expressions exist over spans of time. They are occurrents. Their representations are continuants. No two occurrents can ever be identical. Occurrents are ontologically dependent upon continuants, and the social/legal preference for fixed expressions in IP regimes tracks the simple dichotomy between continuants and occurrents, by which continuants are ontologically primary. There are also some highly pragmatic reasons to prefer fixed, continuant expressions over non-fixed, occurrents, at least when according expressions legal protections. Not least among these reasons is enforceability. If someone reproduces an expression as a continuant, it is much easier to discover the reproduction than if one reproduces an occurrent expression. Infringements are thus easier to locate, and legal action is easier to pursue. Moreover, there is a likely social preference for allowing freedom of actions that makes attempts to stifle occurrent expressions unpalatable.

Would protecting occurrent expressions ever be justified either theoretically or practically, and if so, would it uphold the purposes of IP regimes? One could certainly envision a social/legal system that prohibited reproducing occurrent expressions (inasmuch as they can be *duplicated*) and that required some royalty, license, or acknowledgement for every reproduction. It would require a great deal of mutual trust to work given that these sorts of expressions are necessarily fleeting and difficult to prove in case of some infringement. Moreover, it is unclear whether this would bolster or hinder the purposes of IP regimes, given that such protections might unduly stifle experimentation in expression. It would likely limit forms of expression we feel generally unwilling to hinder, such as free speech and movements. If one faces the risk of unknowing infringement of another's expressions whenever one engages in dialogue, or dances, improvises, or brainstorms out loud, then values that we favor even over those embodied in IP will be threatened. While it is logically possible, devising institutions that would protect occurrent expressions is both pragmatically difficult to conceive and potentially threatening to both the purposes of IP law and other more basic values.

IP Challenges in Present and Future Nanowares

The resulting products from both present and future nanoware technologies will be continuant expressions. The distinction lies in the manner of delivery, because until recently, the author, inventor, or some intermediary has either printed or assembled the final product that gets sold. Enforcement of the first sale right to royalties has been more or less straightforward. Infringing products could be spotted by tracing their point of origin. Counterfeits could be tracked by unique IDs, trademarks, or other materials. Of course, counterfeit products still abound, but the chain of custody of each authentic item, whether a film, book, or handbag, leads to some author, inventor, or authorized intermediary who constructed the authentic product and placed it into the stream of commerce. But all of this becomes complicated if consumers become the producers, assembling the final product themselves. This scenario poses theoretical and practical problems relating the theory of artifacts discussed above.

IP protection has centered around granting monopoly rights to authors and inventors for each first sale of the objects of their authorship or invention. Both present and future nano/microfab products will be assembled more or less by the consumer or at least at the consumer end of the supply chain. So what will constitute the first sale? It will be something very much like, but not exactly, a 'type' rather than a token which will be sold in the scenarios envisioned, or the token sold will at least differ significantly in kind from the token that is ultimately used. While the software that will actually be exchanged, that drives the local fabrication of the eventual object constructed, will be purchased, its purpose will be the fabrication of something else. There is a problem with this that confounds ordinary IP laws and their reach. Namely, instructions, directions, recipes, and other *utilitarian* works of authorship (rather than aesthetic) have typically been *excluded* from most IP protection. The justification has been that protection should only issue for original works of creativity or inventiveness, and recipes, instructions, and directions constitute facts or descriptions of states of affairs, even if listed in a particular order. Software seems to get around this by being both the process and the product. The instructions bound up with software inform the computer to direct the software's own functioning. But this is cheating. It really points to the erroneous original distinction between utilitarian and aesthetic expressions. Software consists of instructions that are performed by a machine. But software also consists of the instructions immanent in the machine while the software is running. All of which is what seems to complicate the application of traditional IP law to software, and eventually nanowares. We have resolved the confusion in software by simply stating by fiat that copyright (and, strangely, patent, which used to be mutually exclusive with copyright) shall apply to software in all its forms. This more or less gets it right, treating the object as it is, an expression, plain and simple. But software is distributed on disks, on memory cards, or in files, the reproductions of which are all identical to one another in the arrangement of their bits. It is that string of bits that gets the protection. Each first sale of that string of bits deserves a royalty because it is that string that is the end product of the sale. What of manufactured items whose manufacture occurs not before but after the purchase, and at the site of the consumer, not by the author or inventor?

Traditional IP will not be able to be applied to the nanowares. It will utterly fail to mesh with IP's aims or methods, or to protect and incentivize in converging technologies. This is partly because traditional IP prohibits reproduction of the object purchased. This has been a sticking point with software which has been resolved by specific exceptions noted explicitly in software licenses (such as the necessary copying and reproduction associated with moving software into the hardware). But in the case of present and future nanowares, the item that will ultimately be constructed is the primary reproduction consumed. The package of software that is used to construct the final object will not be the primary object to be consumed. Rather, it is the chair, RepRap machine, or other widget assembled either by hand, by robot, or by molecular assembly that must ultimately receive protection, even though it is *transferred* to the consumer in some other form, and assembled or otherwise reproduced by the consumer. Both patent and copyright prohibit the sort of reproduction that will actually be necessary for the full use of the widget sold, and to make the promise of present and future nanowares complete. So what will be sold, what sorts of protections can be used to both encourage innovation and make full use of the technology envisioned, and how will we have to adapt our institutions to accommodate new modes of innovation, distribution, and assembly?

There is no natural basis by which we can exert control over ideas. The types are abstract, they are reproduced in minds, and until they are expressed in some way, there is neither any way to determine who holds an idea nor to prevent that conception. There is also only one known way to exert any natural control over expressions, and that is secrecy. If you wish to prevent others from expressing 'your' ideas, you can only keep secret those ideas and hope that others don't conceive of them as well. Once expressed, ideas can become easily reproduced in other minds and then other media. The problem with secret-keeping is that it discourages rather than encourages innovation, so IP regimes seek to promote disclosure and grant limited monopoly rights to ensure profits. IP regimes are entirely the creatures of positive lawmaking, and we are free to create them as we wish, consistent with other important values and rights. Can we conceive of a better way to encourage innovation, help ensure profits, and benefit consumers? Can we develop new institutions that will fit new modes of manufacture and distribution? We can. Some of these new institutions are already developing at the grassroots, and could likely be adapted to fit the world of atoms in the nanowares future envisioned.

Contract versus Monopoly

Because neither copyright nor patent has satisfied a number of consumers and producers of new media (such as software), alternate forms of protection have developed as viable methods of distributing products without government-sponsored monopolies. Beginning with the free software movement that launched the GNU/GPL (GNU General Public License), numerous new, grassroots innovations in contractual agreements have arisen to protect and encourage authors and inventors while enabling greater flexibility and fair use of new works. These innovations recognize the vital role that agreements and incentives play in encouraging innovation, as well as the moral choice involved when inventors and authors seek some compensation for their works. But they also recognize that these goals and choices are not constrained by the models employed historically in the rather recent inventions of copyright and patent laws. Sometimes, new forms of products or services deserve new forms of incentives. In the case of software, a number of the new forms of protection that have emerged have become favored by developers and consumers alike, and have enabled profits to be reaped outside of the traditional, government-sponsored monopolies of copyright and patent.

In general, alternate forms of protection involve various licenses, including shareware, creative commons, open source, copyleft, and others. Underlying all of these various devices is their flexibility and their openness. They are created to encourage dissemination, while retaining either some information flow back to the author/inventor, or some payment, or both. In each case, alternate forms of recognition of authorship or inventor status confer some limited possessory and use rights to the consumer, and retain some rights to the author or inventor. Defining the nature of those rights is the providence of a private contract. So far, most of these alternate forms of licensing have concerned software, and they are entered into through various mechanisms. 'Shrink-wrap' agreements make the license binding when the software is either unpackaged or installed. More commonly, downloading software under one of these schemes involves agreeing to some form of End User License Agreement. There are important issues involved with the propriety of contract in many of these cases due to the fact that many end users enter into license agreements rather casually, and without any of the requisite 'meeting of the minds' generally necessary in contract law for a contract to be legal and binding. We might well ask, however, if a valid and enforceable alternative licensing scheme could be applied to present or future nanowares?

There is no doubt that any form of licensing of products distributed with this technology will differ in important ways from software that is currently licensed through alternate schemes. There are critical distinctions between what is actually going to be exchanged under the sorts of present and future nanowares infrastructures that are conceivable. Unlike an ordinary purchase from, let's say, a furniture store, no material good (of the ordinary kind) will be exchanged. The only material exchanged will be the design templates, CAD drawings, or similar artifacts. The final construction of the ultimate product will be made by the consumer, mediated through some set of machines. So what will be sold? What will be the responsibilities of the vendor and what liabilities will the consumer assume? To illustrate graphically why all of this is so vitally important, consider what might happen should an inventor distribute his or her design, marketed as a new and improved safety helmet for biking, which when constructed by the consumer turns out to cause more injuries than it prevents? Ordinarily, product liability is strict, and flows from the manufacturer to every good manufactured, put into the stream of commerce, and used as it is intended to be used. But in the scenario envisioned, who is the manufacturer? Is the 'publishing' of the design specifications, no matter how detailed, enough to make the inventor potentially liable? To whom will the inventor be liable? Will *anyone* manufacturing the item, no matter how they came into possession of the design (e.g. even if they pirated the specs), be in some sort of liability/contractual relationship with the inventor?

If licensing schemes for physical products are to achieve the promise and potential they are achieving for certain software goods, they will need to be different, perhaps radically different, from alternative licensing for software. In the case of software, the good delivered is the final good, and the range of liabilities is necessarily limited by the medium. So far, software failures are generally incapable of causing physical harms. In the case of physical goods, delivered *via* software, the range of liabilities is greater, and mechanisms for maintaining information and control over each copy are made more complicated. There is no theoretical reason, however, why a license could not be developed that could accommodate physical products. Perhaps the seeds of such a license lie in currently used and successful alternative licenses. Can we adapt any of these to serve the various goals typically associated with IP regimes so that products can begin to be marketed, adapted, and adopted by paying consumers as open source software or shareware currently is?

Some Initial Requirements

Before we successfully apply flexible forms of licensing to physical products, a minimal set of requirements must be established. Among these are the purposes of IP law, namely, encouraging innovation and ensuring that useful information eventually flows into the public domain and to the public benefit. A further, more complicated set of requirements revolves around potential liabilities, including for economic and personal harms. Software so far hasn't had a significant potential to physically harm people (War Games scenarios aside), but physical products distributed first as software and then constructed by the consumer pose real dangers. If a licensing scheme is to be successful in the realm of physical goods, the potential liabilities, and means to address them realistically, meaningfully, and efficiently, must all be worked out before the first big harm occurs.

Alternative schemes work best when they are flexible, and contracts work best when they are plain and easily understood by all parties. In the spirit of do it yourself (DIY) culture, a part of which drives the current trends in nanowares, it is worth trying to devise licensing schemes that can easily be adapted to the particular good sold, and that are based on plain, easy to understand terms, concepts, and interchangeable parts. Legalese will not do. Underlying the success of alternate forms of software licensing is the basic value of trust between consumer and vendor. Openness and free exchange of information between the parties have helped improve the products produced and distributed under these schemes, so that names like Firefox, Apache, and Linux are trusted as reliable purveyors of well-built goods. Alternative forms of licensing should help ensure that this spirit remains alive in present and future nanowares, and trust must remain a cornerstone of transactions. Freedom and openness should be preserved.

There are alternatives. Future nanotech and even present schemes of microfabrication could be modeled after patent and copyright. However, it is hard to *fully* envision. Problems of enforceability, and realizing the full potential of these technologies are two of the largest impediments to fitting nanowares into the straightjackets of patent and copyright. A new form of legal protection might be devised, granting some other form of monopoly right, utilizing a top-down approach. Authors and inventors might prefer the reliability and safeguards seemingly afforded by government-sponsored monopolies to those that might be constructed from the grassroots in a free market. If so, then so be it, but a free market demands that experimentation and alternatives exist, and battle it out for market share, before settling on a particular form of protection over any other. Copyright and patent already exist, and some software authors use one or the other or both depending upon their preferences. Others use copyleft, creative commons, open source, or shareware to enter the marketplace, and some have achieved significant success.

There is already good reason to believe that physical goods and software can theoretically be treated as a kind. Each is an expression – the manifestation of an intentional idea from a mind into the world outside that mind. But because transactions of goods under the sorts of technological infrastructures envisioned and ultimately demanded by present and future nanowares will differ from sales of physical goods with which we have become accustomed, new models for a future supply chain, and delivering nanotech goods, might be better devised and employed. It is worth beginning the task of creating these new models, and testing them on existing, nascent forms of microfabrication such as RepRap Darwin, Fab@Home, and Fab Lab.

Empirical Work to be Done

A proof of concept can easily be devised. It is being devised by some who have been developing versions of the Creative Commons license that will apply to physical goods. The requirements described above can be fleshed out, and existing licensing schemes can be adapted to tangible goods. Early adopters are already tinkering with the technology, improving it steadily, and waiting for the killer app that will help it move from garages to the mainstream. Economic downturns like that of 2008–10 provide a good excuse to move the technology, and the social–legal infrastructure forward. Downturns are only made worse by the backlog of goods that ordinary manufacturing supply chains require. The system clogs up with old, warehoused products, requiring manufacturing slowdowns and sending prices into deflationary spirals. If products were delivered at the point of sale, in a 'just-in-time' manner, then it seems likely that downturns would not be so deep. Moreover, surpluses, waste materials, and recent concerns about environmental consequences of large-scale industrialization also suggest that our old modes of manufacturing need reevaluating. Now is a good time to invest time and energy not only in the technology and infrastructure, but also the institutions that will be required for the full promise of these new technologies to be realized. Those who are tinkering with the technology and seeking funding to perfect it would also do well to work with those of us who are interested in ensuring that nanowares fulfill their ultimate promise, profit those who have invested time, money, and energy into realizing the technology's full potential, and provide the backbone for a new technological and commercial infrastructure that will help turn human creativity loose in the world of physical goods as it has done so in the world of software.

Nina Paley, *Mimi & Eunice* (cc) creative commons

CHAPTER SIX

Authorship and Artifacts:
Remaking IP Law for Future Objects[1]

Rights to Expressions: History and Theory

The development of the legal-institutional notion of intellectual property (IP) is rather recent, historically speaking. It arose in the western world in various forms, beginning in the mid-1500s and early 1600s. In both Italy and Britain, monarchs began granting 'letters patent' to individuals who might improve the local useful arts (by invention) or help to grow their nation's coffers in other ways. One of these ways was, ironically, through piracy. As Britain tussled with Spain over the riches of the New World, the British crown encouraged buccaneers to work against the Spanish, becoming *legitimate* pirates (renamed 'privateers'), through the grant of 'letters patent'. These letters from the monarch gave privateers exclusive rights to a portion of the proceeds from their raids on Spanish ships, forts, and colonies. A sovereign's grant of a monopoly right was a strong incentive also to publish certain works, and to encourage the immigration of creative, inventive persons and their inventions or other creative works.[2]

Before sovereigns invented letters patent, and then eventually copyrights, there was simply no natural way for an author or inventor to protect the fruits of his or her inventiveness or creativity against economic exploitation by any other diligent copier. If one had a great new idea, and wished to profit from it without others copying and depriving one of direct profits, then one had two options: secretiveness or obscurity. Obscurity meant devising inventions in ways that could not be readily reverse-engineered, and secretiveness meant practicing one's art away from the public eye. Either meant risking a great deal. Secrets could be spilled, or independently discovered, and obscurity could be punctured with a bit of knowledge and work. Ultimately, the world of innovation before IP emerged included every possible attempt to keep the lid on new knowledge as much as possible, and ran counter to the scientific trend of the Renaissance and Enlightenment.[3]

Scientific discovery requires openness. It is only by independent confirmation of scientific results by other researchers that general knowledge advances. As the scientific method developed, and institutions of science grew (including scientific colloquia, journals, and other modern elements of the scientific process), the value of openness replaced secretiveness in the realm of discovery and invention. In the nineteenth and twentieth centuries, IP laws often worked in concert with the development of science and industry to promote the transfer of knowledge between the two, and encouraged economic growth and innovation without reliance upon secrecy, which ultimately hinders science. The development of those laws captures some fascinating relationships among people, intentions, artifacts, and related objects. There are some clear trends, some discoveries of first principles, and some questionable applications that bear discussion. All of these provide an interesting background for discussing the nature of the objects involved, and considering the theoretical and practical consequences of applying new potential models for protecting or otherwise encouraging innovation. Let's first look at the categories of being discovered in the development of IP laws, and then consider whether they might be at all based upon first principles or whether they are all just pragmatic categories.

The law has distinguished primarily between those things that are patentable and those that are copyrightable. Patentable objects include new, useful, non-obvious inventions. They may be either products, compositions of matter, or processes. These sorts of things are considered primarily utilitarian. On the other hand, works whose uses are primarily aesthetic have been granted copyrights. I have argued extensively that nothing 'grounds' IP rights, not in the way that real property (land) or chattels (goods – objects that may be physically possessed to the exclusion of others) are grounded in the brute facts of possession.[4] Simply put, while the token you make of your idea (a novel or a steam engine) is your property, the *idea* (the type) can be copied by anyone sufficiently patient or skilled in the art upon which the token you made is based. If your neighbor makes his own steam engine, you have not been deprived of anything, nothing was taken from you leaving you with less than you had. It is this intangible, easily spread nature of ideas that impelled states to begin creating IP rights.

In order to encourage innovation, and to provide an institutional bulwark against the natural inability of inventors and authors to exclude others from copying or using their works, states devised and enforced IP regimes, including the modern forms of copyright and patent. These rights to 'intellectual property' are utterly distinct from traditional forms of property law developed both through common law and positive enactments. While most property rests upon rights to possession and exclusion of others from the things (property) possessed, IP rights involve no possessory right to any particular object. Rather, an IP holder has an *exclusionary* right over the reproduction or first sale of any token instantiating the type he has created. Thus, the author of a novel can forbid others from selling a copy of his story without paying him royalties (at least the first sale of the particular token expressing his story – used books and other media may be sold without paying royalties). The inventor of a new device or process may forbid others from selling tokens of that device, or employing the process he invented (again, the first sale of the object only) without paying license fees. IP diverges from ordinary modes of ownership in that it depends upon some underlying notions of types and tokens, as well as distinctions among nature, abstract entities, and artifacts, all of which are not just philosophically interesting, but practically complicated with the arrival of new types of expressions.

Recently, with the emergence of information and communication technologies (ICT – computers, internet, etc.), the law of IP has proven to be difficult to apply, with debates arising as to whether software can be patented or whether it should be copyrighted. Both have been done, a practice that challenges the previously mutual exclusivity of the two prevalent categories of IP. Even more recently, patenting of unmodified genes has raised further questions about the adequacy of IP to deal with emerging technologies and modes of innovation.[5]

The law of IP offers us a rich guide for understanding the nature of the underlying objects of *intellectual property* themselves. There are some long-standing categories in the law that reflect sensible distinctions among types of objects, even while there are some that may also be suspect. Let us consider some of these categories, based as they are on historical distinctions, and return to look critically on where the law might diverge from its own foundational rationales.

Nature, Creation, Artifact, and Invention

There have always been limits to what may be patentable or subject to copyright. One of the first hurdles to IP protection is originality.[6] In order to be patentable or copyrightable, one must have devised something that did not already exist. This means it cannot have been the work of another, nor could it be something merely natural. Ordinarily, this is uncomplicated. Borderline cases have arisen rather more frequently recently that sometimes challenge assumptions about creativity and inventiveness or highlight the rationale for this distinction. In the 1970s and 1980s, patenting algorithms as parts of computer software raised issues on the patent eligibility of mathematical formulas (discussed below), and more recently the distinction between the natural and the patentable has arisen in the debate over patents on genes. All patent statutes, however, hinge upon some measure of human inventiveness being brought to bear on nature – the creation of something that previously never existed.

There is then a class of objects that we have decided should never be protected by IP, even though we often consider it to be the domain of human conception. Science is focused upon discovering 'laws of nature' and describing them accurately. But there is much technology that depends upon scientific discovery. Descriptions of laws of nature, or theories, are not patentable although they are human inventions. To the extent that $F = ma$ describes a natural relation between force, mass, and acceleration (Newton's second law), it is nonetheless a description, not the 'law' itself. In other words, force, mass, and acceleration exist, and preexisted Newton, relating as they do with or without human intention. Newton's second law remains a human invention, describing a natural state of affairs. '$F = ma$' is, technically, an expression. It is a very useful expression, a statement describing natural relations that are universally true (setting aside for the moment questions raised by both Einsteinian relativity and quantum mechanics). So, while it is clear that the natural relations between force, mass, and acceleration cannot be the subject of IP law (as they are not *new*), why can't the law *as formulated* be granted IP protection? Could Newton have patented or copyrighted '$F = ma$' given that it was a new formulation, conceived by a human, of something which is, after all, quite useful?

No current IP regime would allow Newton to patent his laws of thermodynamics, nor would doing so suit any of the purposes of IP law. Although expressions of scientific laws are the result of human creativity, and they are provisional models of natural phenomena rather than the natural phenomena themselves, granting IP protection to Newton's laws, or relativity, or other depictions of natural phenomena, would inhibit invention rather than promote it. It would contravene the practical purposes of IP law, and it would defy sense, and would be akin to granting Joseph Priestley a patent to O_2 as discussed above. It would also be a very limited protection, guaranteeing only that others could not use the formulation of that law of nature (which was a new expression), but not preventing anyone from *employing* that law given it is as old as the universe itself. There is a distinction between the process of science and that of invention. Science seeks to discover the laws that underlie natural phenomena; the purposes of science include improving understanding, prediction, and control over nature. Invention is often a result of science, but science is often not necessary for invention, nor is invention the necessary consequence of science. The objects of science and of technology are also different from their processes. While the depictions of natural laws are human constructs, for instance the languages of mathematics or of physics and their theorems and hypotheses, the goals of these sciences are to uncover truths about nature, resulting in more accurate descriptions, fuller understanding, more useful predictions, and better control.

Even if nature itself could be owned in any conceivable manner, extending control over its laws or products would not likely encourage innovation or authorship. The first essential distinction inherent in IP law between nature and invention is based upon sound division between those things that exist without human intervention and those that require it. The laws of thermodynamics exist with or without our depictions of those laws through languages or symbols. Depictions of those laws might be human creations, but their usefulness depends upon their hewing as closely as possible to the facts of nature. There will thus be a limited number of ways of depicting those laws usefully, unless we choose to multiply our languages of depiction, creating new physics or mathematics, or biological lexicons. But doing this would only impede innovation and hinder science.

In order to maximally encourage both science and innovation, there is a balance struck in the law of IP that prohibits protection of mere depictions of natural laws (the domain of science) and allows protection typically only for *applications* of those laws through some new technique: some artifact or inventive process. IP laws apply, based upon the pragmatic purposes of IP regimes, only to expressions that are new and useful, and that are not depictions of natural laws. We can see the usefulness of this distinction if we consider the difference between mathematical formulas and 'algorithms'. A formula expresses an equivalence that is part of nature. The formula for the area of a circle, πr^2, represents a truth about nature, and its discovery is generally attributed to the ancient Greeks. It depicts nothing new, but rather a truth that has always been a part of nature, even before the abstract entities of mathematics were discovered and the languages of mathematics invented. Algorithms, however, are inventive. An algorithm is a description of a method for solving a problem. Thus, the following is an algorithm for determining the area of a circle: (1) measure the radius of the circle, (2) multiply the radius by itself, (3) multiply the result by π. The formula for the area of a circle represents an abstract entity – a type inherent in nature and not the result of human invention, an idealized mathematical construct that has no empirical counterpart in nature. It cannot be owned under any permutation of IP law. On the other hand, the algorithm described above is a process that can be applied to achieve desired results; although it relies upon laws of nature, it is not itself a law of nature. Under some interpretations of IP laws, particularly current US patent law, the algorithm might be patentable because of some ambiguity in the courts' interpretations of patent laws and a lack of clarity regarding the underlying objects at issue.

In a famous trilogy of cases decided within a ten-year span, the US Supreme Court was forced to consider the patentability of various algorithms due to the emergence of software and a scramble for patents over computer programs. Beginning in 1978 with *Gottschalk v. Benson*,[7] followed by *Parker v. Flook* in 1978,[8] and concluding with *Diamond v. Diehr*, in 1981, the US Supreme Court grappled with the question of when an algorithm should be patentable as opposed to when it may not be because it constitutes 'prior art'. In *Diamond v. Diehr*, which remains the prevailing law in the United States, a process that made use of a software algorithm was at first rejected by the Patent and Trademark Office (PTO) as ineligible for a patent. The process incorporated the 'Arrhenius equation',

$$E_a = -RT \ln\left(\frac{k}{A}\right),$$

into a process used to cure rubber. The Supreme Court reversed the judgments of the PTO and the lower court, holding that although the invention contained 'prior art', by way of a mathematical formula, the 'invention as a whole' was new. The patent could not foreclose use by others of the mathematical formula, but would prevent others from making or marketing the same invention, the new combination of natural phenomena and materials.[9]

So what is the common denominator to inventive objects that takes them out of the realm of the natural and makes them protectable by IP? Nature must somehow become transformed by human intention and action.

The result of that transformation is an expression of an idea, either by way of some artifact (a continuant) or by way of some action (an occurrent, like a service, dance, speech, etc.). The catch is that for occurrents to garner IP protection, they must somehow become transformed once more. They must become 'fixed' in some 'tangible medium'. This poses a problem for some modern inventions whose dissemination has seemed at times to consist of selling the types rather than the tokens, or performing a service rather than exchanging a tangible good. The confusion between types and tokens, confounded because of the seemingly ephemeral nature of software, is not based upon any real distinction between software and other tokens sold or distributed under existing IP regimes. Each instance of a piece of software is clearly a token, and it is the token that is the subject of each software sale. A piece of software exists, in its saleable form, in a fixed medium. It is properly subject to either patent or copyright (strangely) under current interpretations of IP law. But this odd fact raises questions about the classifications of types of expressions embodied in the dichotomy between copyrightable and patentable objects, and suggests that it may be time to revise our current understanding and application of these categories.

Revising Our Relationships with Artifacts

We can see the emergence of a simple hierarchical taxonomy of IP. At the top is nature, or 'brute facts'. While parts or tokens that are natural (a piece of land, an individual sheep) might be possessed and then owned under some positive institutional scheme, no claim of ownership nor reasonable means of possession of any of *nature's* types could be conceived or exercised. No brute fact manner of enclosing nature or her laws could be created, nor would devising any institutional mode of that kind of ownership promote the goals of IP regimes: innovation, profit, and public benefit. IP regimes are creatures of institutional reality that grant monopoly rights to authors and inventors who *change* nature, who alter it in some *new* way. The objects of IP regimes are exclusionary rights over reproduction or first sale of any token of a certain type. So far, only continuants, created by humans, have been afforded the exclusionary rights of IP laws. A machine, a story, a song, a dance, as long as it becomes 'fixed' in some tangible form (as a continuant) may only be reproduced, sold, or otherwise enter the stream of commerce with some royalty to the authors or inventors. Speeches, dances, songs, services, and other occurrents, though they may be the products of human intention, may not be protected under any current form of IP protection if they do not become, somehow, a continuant.

As with all IP regimes, this is a matter of choice and not necessity, although the choice might be firmly grounded in pragmatism. It is a choice that could well be challenged, as could that behind the artificial distinction between the objects of patent law and those of copyright law. In case distinctions among types of objects fail to promote the underlying goals of IP and its institutions, we should feel free to reconfigure these distinctions as we see fit, to better promote those goals, and as technologies and contexts change. The only restriction to how we might reconfigure these institutions is internal consistency, a criterion by which the current regimes utterly fail (as the examples of software and gene patents reveal).

If we admit that there are good, practical reasons to provide some institutional incentives for authors and inventors to take the risk of realizing their ideas and exposing them to the marketplace, and concede that the current forms of IP that exist lack internal consistency, then we should feel free to alter our institutions. What we cannot alter is the distinction between the natural and the artificial. There is the world of unmodified nature, and there is everything else. The 'everything else' includes all modifications of nature by man (assuming we are not yet interested in creating IP regimes for animals). Novelty seems like a good criterion to preserve if we wish our IP regime to encourage innovation. But we might well question whether the current type/token distinction embodied in IP's idea/expression dichotomy necessarily captures something important, or potentially complicates or inhibits the ways in which we might encourage innovation, especially as we delve more deeply into exchanges of nanowares.

Ideas are certainly distinct from expressions, and ought never to be confused. But the law of IP grants protection to ideas that most IP practitioners deny. It's true that only *expressions* of protected ideas can be prohibited by IP regimes, and that no one may be denied the ability (perhaps founded upon some sort of right of conscience) to think an idea, even if he is not the original author. But the fact is that IP protects ideas by way of prohibiting their expressions in media other than minds. While this is less intrusive than prohibiting the holding of ideas in minds, it is still a significant restriction on speech, if not necessarily thought. It prevents the dissemination of certain ideas by certain routes. IP laws give authors and inventors limited monopolies over types, to the extent that others may not utilize those types in any number of possible ways.

While we generally speak coyly about IP curtailing *only* our relationships with tokens, the net effect is that it curtails our relations with types, except inasmuch as those types may still be comfortably nestled in our minds, if nowhere else.

So what prevents us from reconfiguring some future form of IP protection to encompass *types* proper, rather than the somewhat awkward and misleading manner in which they are now supposed only to cover tokens (expressions rather than ideas)? Considering the direction that innovation is going, as software, services, and products begin to merge with the advent of true nanotechnology, the type/token dichotomy as expressed currently in IP law may well be archaic. The present hierarchy of the law of IP looks like Figure 6.1.

Creation and Dissemination of Types versus Goods

It is not surprising that there are some seemingly arbitrary divisions and categories in the current scheme of IP as it is a feature of the positive law, ungrounded in any natural law and driven by purely pragmatic concerns. Why extend protection only to continuants and not, for instance, to a performance or a service? What is the distinction between 'inventions' and 'works of authorship'? Is there reason to be flexible in light of emerging technologies and trends in existing technologies, and to reconsider this hierarchy and its current divisions? Moreover, might we reconsider the manners in which we choose to create our IP schemes? There is some reason to think we should.

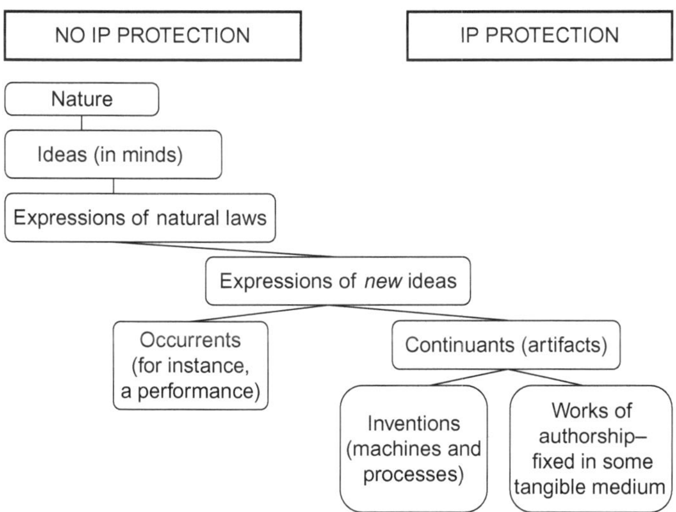

Figure 6.1 The Current Ontology of IP Law

The emergence of ICT and the coming age of nanotechnology ought to continue to undermine the traditional distinctions inherent in the current law of IP. Things that we have considered to be 'fixed' objects will become programmable, fluid combinations of software and flexible materials. As the division between software and hardware breaks down, as did that between written works and machines through the development of software, patent and copyright will prove to be archaic modes of protection or encouragement for innovation. For instance, when innovators are no longer involved in disseminating physical goods, patents may *hinder* the creation and dissemination of new, nanotech-driven *types* given that a patent would technically prohibit the 'reproduction' (production, really) of the good by the consumer. Similar problems have hounded software patents and copyrights as multiple reproductions of the software are typically involved in each sale, as part of necessary installation procedures and archival backups. So far, IP law has worked through these technical impediments by developing kludges of sorts, exceptions that allow for what would otherwise be forbidden. The constant need for new kludges as technology evolves suggests, however, that we might need a completely revised framework and our old system ought to be scrapped. We can assume that the goal of IP remains the same: encouraging innovation and moving the products of ideas (eventually) into the public domain. These goals emerged out of the historical desire to promote innovation and prevent secrecy. But recently, secrecy has become a mere technical hurdle to overcome, and modes of innovation have changed. No longer do many innovations require large capital investments, and the tools for creating and disseminating new goods have become more accessible given both ICT and nanotechnology's precursor: localized fabrication.

Just as the emergence of the PC allowed 'basement programmers' to create new and profitable goods with little capital investment, so too will computer-aided design, localized manufacturing, and eventually nanotechnology and desktop manufacturing allow for innovators to realize and distribute new physical goods without the traditional capital burdens of factories and physical distribution networks. But IP laws remain a significant impediment to this scenario for the reasons outlined above. The categories we are used to will not work. Innovation and dissemination will no longer be of the goods themselves, but of the types. Current IP regimes do not account for this sort of innovation, production, or distribution, and as we have seen with the wholesale violation of software copyrights and patents through grassroots peer-to-peer (P2P) networks, innovators will be foolish to disseminate their creations without better, more flexible, and enforceable institutions and laws that take into account what will actually be created and sold. Consumers too will be wary of purchasing goods where the rights they might have to use those goods are ill-defined, and where the law might have made accommodation through yet another legal kludge, rather than through a profound reconsideration of the objects and relationships truly involved.

Setting aside the categories of patent and copyright, and their inherent shortcomings, can the taxonomy of types in Figure 6.1 guide a revised outline of IP protection? If we accept that converging technologies will continue to undermine divisions between types of expressions, could a unified scheme, covering all potential expressions, be employed in place of current copyright and patent laws? I proposed one such scheme in *The Ontology of Cyberspace*, in which I suggested a single system of protection, much like copyright (but with shorter terms of protection), that could be extended to all expressions (including inventions). There is no theoretical reason to think we couldn't pursue some alternative, although devising a workable system from scratch seems daunting.

In fact, numerous alternate schemes to traditional IP protection have already emerged through the development of licensing systems, moving the means of protection into the hands of producers and consumers, and avoiding the institutional costs of formal registration of patent or copyright. If states are to remain involved, then it seems that in order to encourage maximal innovation and reduce barriers to entry, lightweight schemes like copyright (which involves no complex or costly review process, and only a very inexpensive registration for *prima facie* proof of first authorship) could replace the unnecessary dual copyright/patent system now employed. Our decisions should be guided by both the practical goals behind the institutions of IP and the underlying ontology, which is founded upon some well-deserved distinctions between ideas and expressions.

Ultimately, we could envision an economy of *types* rather than goods. In some ways, the software industry already functions this way. With increasing frequency, consumers download software rather than purchase it burned already on a CD or DVD. What they are purchasing, often by accepting a complex End User License Agreement (EULA), is a right to reproduce the type as an instance on their PC. What is transferred is not the program in its final form, but rather the right to reproduce some intermediary form, and then to unpack and install the program in a final useful form. A unified treatment of all expressions, and licensing systems, depending on private arrangements between consumers and producers is already working as a model for alternate approaches to disseminating nanowares. These schemes treat the transaction as an exchange not of the final token, but of a right to the use of a type through various intermediate tokens, in ways that current IP regimes cannot. The current copyright system also serves as a model for this sort of exchange, as the sale of a book transfers most importantly certain rights to the type inherent in the token. In the future, the marketplace for all goods will look more like this, where types are the primary unit of exchange, their values and prices unencumbered by costs associated with mass production and capitalization, and parties to exchanges will be free to work out the details of the rights involved in each new exchange.

Ethical Problems with Traditional IP

We have seen above how the law of IP has been warped, and its principles extended beyond reason. We have delved a bit into the practical problems posed by attempts at applying the current IP scheme to nanowares. But there is a fundamental moral argument that can be made against the very notion of IP, and that helps to argue for a new conception that will be outlined in the subsequent chapters. Simply put, IP draws an unwarranted and *anti-natural* relation between authors (or inventors) and artifacts. Patents and copyrights extend protection over things that *ought* not to be protected, if we are to take seriously a number of basic assumptions we otherwise hold regarding freedom of expression, freedom of conscience, and rights to tangible property. Let's examine how IP law might not just be inefficient, impossible to apply to a future of nanowares, and costly, but also unethical *per se*.

AXIOM 1: *We have fundamental rights to autonomy over our bodies and our minds.*

I take this axiom as uncontroversial in western, liberal democracies. Liberalism presupposes that we are by nature entitled to certain freedoms, and among these are freedoms over our bodies, minds, and our property. This principle is the cornerstone of modern democratic republics, and can be traced back to Greece, as well as Enlightenment Europe.

The right to autonomy involves obligations for others to not interfere with our bodies, our minds, and our property. Individual sovereignty means that we can dispose of our bodies as we wish, direct our thoughts in any way we please, and hold our property without interference. Since Locke and later Mill, this liberty entails the peaceful enjoyment of these rights against not only others but also the state. The limit of our liberty is where our acts interfere with the liberty of another, as expressed by John Stuart Mill's 'harm principle' or 'principle of liberty'.

AXIOM 2: *We have fundamental rights to freedom of expression consistent with Mill's harm principle.*

This is partially implied by Axiom 1, our expressions are the results of the exercise of both our bodily and mental autonomy. An expression is the making-manifest of ideas into the world outside of minds. Speech acts and artifacts are the primary means of expression, although physical acts are also expressions. Limits to free expression typically center around the harms caused by them, with laws that prohibit certain expressions being generally disfavored. Instead, laws regarding limits to expressions typically focus on punishment *after the fact* for harms that have been caused. 'Prior restraint' of expression is generally disfavored. Which brings us to our first premise:

PREMISE 1: *All man-made objects intentionally produced are expressions, and IP law involves monopolies over expressions.*

IP rights are not positive but negative. They are exclusionary. The holder of a patent or a copyright has the right to exclude others from doing certain things with types. The objects of all IP law are both the ideas (or types) *and* the expressions (tokens). This is so because IP laws limit what others may express, and give owners of copyrights and patents the exclusive right to make and sell their expressions. Exclusive control of expressions restricts what one may do with types. Although anyone can *consider* any type, *think* of it in any way, *muse* about it, or even *discuss* it (as long as one doesn't verbatim reiterate an aesthetic expression in some fixed medium), protected types cannot be expressed by others. A machine is an expression, as is a novel, in that each manifests some idea into the world beyond the mind that conceives it. One is primarily utilitarian, and the other is primarily aesthetic. In each case, the holder of the IP right can prevent expression of the same idea in the same or substantially similar way by anyone else. IP is thus a limit on expression unlike other limits on expressions in that it limits expressions without consideration of harm or potential harm.

The rights conferred by IP are not really property rights, but they interfere with property rights. The owner of a patent cannot claim possession of all instances of his or her patented type, but can limit what is done with each of those instances, as well as what is done (e.g. by unauthorized expression) with the type. IP limits what one may do with one's body (by expression) and one's property. Are such limits aligned with the ethical bases for autonomy, and for rights to other forms of property, or are they contrary?

PREMISE 2: *Rights of ownership are grounded in brute facts of possession, and laws are just when they are grounded (and not contrary to) brute facts.*

Property law is rooted in the same principles that ground Axiom 1. Autonomy in one's self, which I take to be a primary and fundamental right, means the obligation by others to leave us undisturbed in the peaceful enjoyment of the space we occupy. This means we are free to occupy a certain area of space, and hold what we possess, without others attempting to dispossess us of either our place or possessions, unless some prior claim to either exists. The justice of property law is grounded in the brute facts of self-possession and possession of material goods.[10] We establish just ownership pre-legally, by the peaceful occupation (and improvement or use) of land and movables as long as we do not dispossess others in the process of their just ownership. This is a more or less Lockean conception of property. Property acquired by these means is acquired justly and neither our neighbors nor the state can justly deprive us of our justly acquired property.

Laws that deprive us of our property are thus unjust, and laws that recognize the grounded nature of just rights to possession are just. IP laws appear not to be related to brute facts at all. IP rights confer the right to exclude others from making reproductions of our expressions, they thus foreclose the use of a type by others, and confer no possessory rights over tokens. Ordinary property rights regard tokens and not types. Because there is no natural way to possess types to the exclusion of others, since a single idea can be held by many without deprivation or trespass, IP laws were created to meet the perceived pragmatic need to encourage innovation *and* disclosure of innovations. These laws are recent, and not grounded in any brute facts. They are instead a *reaction* to the brute fact that ideas are non-rivalrous and non-exclusive, and thus freely available to all to capitalize upon once expressed.

Positive (man-made) laws (that are *just*) regarding property are founded upon *natural law*. IP law is not founded upon any natural law or brute facts. While some argue that authors have a *moral* right to the exclusive use of their expressions (and this point of view dates back about 200 years in French law in particular) most modern theorists of IP accept that there is no natural basis to claim that anyone has a right to exclusive use of one's expression *types* once expressed. If we base our moral claims on the notion of harms, as we do with other natural rights and obligations, it is difficult to support the notion that one who appropriates a type in any way causes the originator of that type any conceivable harm.

PREMISE 3: *There are parts of the world that are not susceptible to ownership.*

Both property and IP law already recognize this premise to an extent. Namely, there are portions of geophysical space over which no valid claim of ownership can be made, and there are parts of the universe about which no valid IP claim could be made. Just ownership, both as a brute fact and as an institutional one, derives from the exclusive and rivalrous qualities of certain types of things like geophysical places and material things. There exist things that are neither rivalrous nor non-exclusive which are thus not susceptible to either possession or ownership. There is simply no means by which certain physical phenomena like radio spectra, laws of nature, or abstract entities could be possessed. These sorts of objects cannot be enclosed logically. There are also parts of the universe that cannot be possessed *practically*. Property law recognizes that certain parts of the world, even geophysical places, cannot be enclosed to any *practical* extent, and thus excludes them from ownership claims. The law has established a category, based upon ancient monarchial practices, called the 'commons'. The commons were originally geophysical places (land) that the sovereign set aside for those with no means or land of their own to use to eke out a living. But we now can note the existence of other types of commons. The categories of commons are commons by choice and commons by necessity. Commons by necessity likewise fall into two categories: that which exists by *logical* necessity, and that which exists by *practical* necessity.

Commons by choice are places or parts of the universe that could be possessed to the exclusion of others, and thus over which legitimate claims of ownership *could* logically exist but for the choice by a sovereign, or preferably by a community, to set aside some space or thing for common usage. This is the ancient form of commons recognized above, and existing since at least medieval European sovereigns began the practice. Commons by practical necessity have also existed for some time in the form of places that no sovereign could practically claim, and thus that could not be granted to subjects or citizens as legitimately owned. International waters are an example. Although one might be able now to mark with GIS some boundary for some part of the surface, or even three-dimensional space within an ocean, prior to GPS technology, marking out, defending, and thus properly possessing such a space was a practical impossibility. It remains quite difficult even now, though it is certainly now logically (and technically) possible. Radio spectra, sunlight, and even parts of outer space or the upper atmosphere also fall into this category. Each of these could be possessed in some logically conceivable way (tougher, though, with radio spectra, unless one is certain one will always have the biggest power source), and yet possession and control of each of these is practically extremely difficult.

Commons by logical necessity are those parts of the universe over which no exclusive possession could logically occur, and thus over which no grounded property claim could exist. Abstract entities (like *pi* or infinity), laws of nature, and naturally occurring phenomena and products all exist in ways that cannot be enclosed to the exclusion of others. No one could logically claim exclusive possession of abstract entities, and the 'possession' of one does nothing (and can do nothing) to dispossess others of its use. We all exist and enjoy the world, including our various freedoms, due to free, non-exclusive and non-rivalrous phenomena like gravity, evolution, the way that photosynthesis produces oxygen, gravity's effect on objects, etc. Moreover, attempts to curtail the use of these sorts of commons by others are attempts to exclude them from use and enjoyment of something that not only cannot logically be curtailed, but also the use and enjoyment of which are free and necessary to all. Thus, commons by logical necessity not only cannot be monopolized, but *ought* not to be attempted to be monopolized due to the necessary harms associated with such an attempt.

PREMISE 4: *Ideas belong to the commons by logical necessity.*

Ideas are abstract entities. Expressions are not. Take any idea, for example, the idea of steam power, or the idea of a steam engine, or even the idea of a particular steam engine. You are now holding thoughts of these ideas in your brain, and the thoughts (for argument's sake) have material qualities. They exist in a certain place, for a certain time, in a certain 'shape' (in that certain neurons fire in specific ways to hold that idea in your thoughts). The ideas themselves lack any material attributes, just as the ratio *pi* lacks any material attributes. Ideas are instantiated in the world in various forms, sometimes by natural phenomena (like otters, rocks, stars), and sometimes by way of artifacts (like expressions, or even as thoughts). There is no manner in which the thinking of an idea can be done to the exclusion of others thinking that idea. Moreover, attempts to curtail others from thinking or expressing an idea are invasions upon basic rights enunciated earlier regarding freedom of conscience and expression.

It is logically inconceivable that one could claim a monopoly right over something like *pi*, or to any other idea. All ideas are abstract entities, even when they become instantiated in some expression. There are any number of possible configurations even of steam engines, and each configuration instantiates a number of ideas, including the idea of steam power, and that of steam engines, and the idea of *that particular* steam engine. IP rights grant monopoly privileges allegedly over particular ideas, but their reach is greater than any one particular apparatus or story. Anything that bears substantial similarities to a particular expression risks being found to violate the monopoly privilege granted under copyright or patent.[11]

IP laws extend to ideas, and prevent their expression by others. This contradicts axioms stated above regarding freedoms of conscience and expression. IP laws are also not grounded in any brute facts nor natural law.

Conclusion

IP laws are not grounded in any natural right to expressions, and impede fundamental rights. IP law is *per se* unethical if we take seriously our rights to freedom of conscience and freedom of expression. There is no natural basis for IP rights. They are monopolies granted as a privilege by sovereigns, a pragmatic attempt to encourage innovation because of the inherently *free* nature of ideas. If a positive law contradicts a natural right, it is the natural right that should prevail, and the positive law that ought to be called into question. This relatively new legal institution is often taken for granted, assumed to be necessary to propel innovation, and yet it appears to conflict with fundamental moral axioms commonly held. Besides having good reason to question the economic efficiency, success, and fair application of IP over time, we now have a *moral* reason to question these institutions. We can move forward now and consider what might replace them, what new institutions could be designed to be more efficient, to better encourage innovation in nanowares, and to be ethical.

Case Study

Publishing Goods

As discussed throughout this text, published expressions and manufactured 'expressions' were once easy to distinguish. A book did not appear to be like a clock, and vice versa. But both clocks and books were sold as goods, and inventors of clocks and authors of novels came to be protected through IP regimes designed to fulfill different perceived needs. In the one case, utilitarian expressions, such as clocks, were afforded patents, and in the other, aesthetic expressions were afforded copyrights. In this text and in others I have argued that the ontological distinctions between aesthetic and utilitarian expressions do not hold, and so a single unified scheme of protection (if we insist upon protection) would suffice for any man-made object intentionally produced. Let's consider how a unified scheme would work, and how clocks and other utilitarian objects could be treated like novels, and whether a copyright-like scheme would accomplish the goals of IP, promote innovation, and help ensure profits for innovators. Finally, we'll apply this logic to nanowares, and consider whether and to what effect a unified regime of IP would work.

When an author or other creator of aesthetic expression creates something new, it is automatically entitled to copyright protection under modern copyright regimes. At one time, that protection lasted a mere fourteen years, after which the work lapsed into the public domain. While protected, the creator enjoyed a monopoly over copies and first sales of his or her works. This monopoly is cheap to get, and valuable to hold, allowing creators of valued works to derive income from their works of creation, and ensuring that the public benefits from both the production of valued works and the movement of those works into the public domain for all to use after some time has passed. Other works are considered to infringe upon the creator's monopoly rights if they are 'substantially similar', and are not proven to be the result of independent creation. Unlike patent law, if two substantially similar works were genuinely created without either creator's knowledge of the other work, copyright holders may then each benefit from their acts of creation. In patent law, the first to get the patent wins, regardless of *bona fide*, innocent independent creation.

Arguably, copyright law has encouraged a great deal of aesthetic creation, ensuring for authors and artists monopoly rights over their works, allowing monopoly pricing while protected, and leaving sufficient room for the creation of similar, though not infringing works. The legal scheme created by copyright law means that no prior review of works is necessary, no expensive filing fees need be paid, and no attorneys need be consulted in order for creators of copyrightable works to be able to immediately begin to reap the rewards of their creativity. In case of dispute, the facts and entitlement to a monopoly under copyright are resolved in court. Over time, the copyright term has increased significantly, and now authors enjoy a monopoly during their entire lifetimes, and after they die their estates may continue to reap monopolistic pricing for an additional seventy years. But as mentioned above, copyright terms used to lapse even before an author died, in terms that were once as short as fourteen years.

Case Study: *Publishing Goods* (continued)

Could a copyright cover a steam engine? There is no reason to think it couldn't. Consider an inventor of a new machine and the creator of a new work of aesthetic expression. Each attempts to devise some unique expression of an idea. The steam engine created, for instance, by Robert Stirling and that created by James Watt, each express the idea of the translation of heat into pressure into mechanical energy. Each does so with a unique arrangement of mechanical parts. The novels *Ulysses* and *Odysseus* similarly represent the general idea of an epic adventure, and in fact *Ulysses* is purposely derivative of *Odysseus*. But the particular arrangement of parts, the characters, unique natures of the locations, etc. make the two works clearly distinct. Both Watt's and Stirling's steam engines have their merits, and each has increased the sum of human creation, and provided means to turn heat into mechanical energy. Consider what might have happened had Watt enjoyed a copyright rather than a patent on his steam engine. The energy spent by Watt in defending his patents, and in preventing competitors from entering 'his' markets, might have been better spent by Watt in improving his own engines, and the delay noted by Boldrin and Levine in the efficient adoption and widespread use of steam power might never have occurred. Watt could still have exacted monopoly rents over the use of his engines, and others could have entered the market with competing engines, as long as theirs were not substantially similar, or could be shown to be the result of independent creation.

For a more modern example, software publishers routinely copyright their expressions (and some also patent them). Comparing the elements of allegedly infringing software and comparing elements of allegedly infringing steam engines are analogous processes. Copyrighting software as expression also takes into account that expression can occur on different levels. Apple's expression of the idea of a graphical user interface differs from that of Microsoft, and that of Linux. Each uses elements that are so general and necessary to accomplishing the goal of allowing users to use a system with a mouse and a pointer, but each has unique elements that make the particular product work in such a way. As well, the underlying algorithms differ. Just as two novels may tell the same general story even while the particular elements of each differ, so too can software accomplish generally recognized useful goals while preserving individuality of creation, and encouraging innovation in the particular arrangement of parts to accomplish the same goal.

Like software, nanowares will (or do) express ideas at different levels. They also do so in multiple configurations. The expression of the full design specs of the final product is itself a unique expression of the idea of the final product, and the final product is also an expression of the ideas inherent in that product. Treating both as a copyrightable whole solves, in many ways, the problems associated with attempting to patent something that requires eventual construction at the user end. A copyright on the full design specs, or other software used to transfer the entire set of instructions for construction of the final product by the user, would cover and prohibit unauthorized reproduction of those specifications. Extending that copyright to the final product, or recognizing that the final product too is a unique expression, affords the creator a monopoly also over the final product without the necessity for a patent search, filing, etc. Those who seek to try to infringe upon the copyright for the product do so at the risk of being sued if their products are substantially similar, and if they fail to prove that their product was the result of innocent, independent creation.

Applying a copyright scheme to nanowares would allow more rapid innovation, as well as reducing costs up front. Part of the value statement for nanowares is reducing the amount necessary for capitalization, and reducing barriers to entry for innovative products in new markets. Some would argue that rampant copying will mean that creators remain unprotected in a unified copyright regime, but liability for infringement upon copyrighted goods has worked for some time in helping to thwart bad-faith copying, and has arguably encouraged production of aesthetic works due to the possibility of monopolistic rewards. A single, copyright scheme may

Case Study: *Publishing Goods* (continued)

well work (but hopefully with significantly reduced time periods of protection) to both encourage and protect innovation in nanowares should we decide that institutional frameworks are necessary. We just need to recognize that there is simply no good reason to continue to pretend that there is a bright line distinction between 'utilitarian' and 'aesthetic' works.

CHAPTER SEVEN

Economics, Surplus, and Justice

We should assume that innovation is good. This assumption is written into the US Constitution's enabling clause for patents and copyright. The purpose is to promote science and the useful arts. Promoting science and the useful arts was considered so important that a source of US federal power embodies this purpose, and the Patent and Copyright Acts that were passed eventually have the imprimatur of Constitutional authority. There is no similar Constitutional power giving the Congress some role in ensuring that, for instance, the hungry get fed. So why is promoting science and the useful arts so important, and giving food to the hungry less so?

Science and the useful arts are vital in a free market based economy, because from these flow the *ability* for those who wish to provide for themselves some means to do so. Those who do not invent or otherwise create progress, nonetheless *benefit* from progress. With new cultivation techniques, and new devices to harvest grains in bulk, came the means to increase crop yields and reduce the cost of grain. With plentitude prices drop. Our modern world of plenty is built upon progress, which is itself built upon the steady, ever-upward march of science. Along with progress in science, which seeks to understand natural phenomena and devise laws to explain them, comes a steady increase in *techniques* or *technologies* that apply those laws to artifacts. Artifacts, which extend human creativity into the world around us, improve our lives. The free market system assumes that the steady march of progress through science and technology improves life generally. In some ways, this is borne out by the evidence of the past 150 years or so, since the dawn of the industrial age. Standards of living throughout the world have improved, if we measure based upon *per capita* income, accumulated wealth, and access to food and medicine. Life spans have similarly increased. Although pockets of deprivation survive, and billions of people still do not share equally in the benefits of progress, it is hard to deny that we are better off now than we were, as individuals and as a species, just 100 years ago. To promote science and the useful arts is to promote the general welfare, to improve the lives of all, though not necessarily equally. The modern world still suffers from tremendous inequities, which nations attempt to solve (or don't) in their own ways. The path we have chosen is based upon the idea that creative people will profit from their innovation by bringing their technology or techniques to market, and the market ensures the survival of the fittest ... in theory. In a free market, ideas compete and the best survive. In a competitive market, prices fall over time as competition sparks efficiencies and improvements, and as demand increases, margins on each individual good may fall, even as profits accumulate.

If we insist on applying a system of intellectual property (IP) rights in order to promote innovation, then we must continue to monitor their effects, even as we challenge also their underlying assumptions. The IP system is itself an enormous bureaucracy, involving hundreds of thousands of people, including examiners and attorneys, as well as special courts and judges. This bureaucracy is costly, as all are, and involves not just immediate costs of administration, but friction in the marketplace. The free flow of ideas and goods is necessarily impeded. From a strictly utilitarian viewpoint, we ought to question whether the return on this investment is worthwhile, both for individual players in the market and for the market as a whole. IP is not a free market device. While free markets must respect property rights, as we have discussed in previous chapters, there is no sense in which IP is a property. Rather, it is a governmentally sponsored limitation on ordinary, natural property and speech rights. The burden is then on those who wish to maintain the IP system to show that it is achieving the ends for which it was established, and not impeding either the economy or justice. Patents create scarcity where there is none. They do this ostensibly because a lack of scarcity will not help ensure that a successful product may recoup the expense and ingenuity involved in devising it, and it is believed that ingenuity ought to be encouraged and rewarded by extending an artificial monopoly to inventors for some period of time. Do they succeed? Is innovation necessarily increased in a patent environment, and is the market enriched? There is an increasing body of evidence that argues against patents as an effective means of either promoting innovation or propelling economic development. Should we embrace such artificial monopolies, even if they were proven to work in promoting innovation, or do underlying concerns of justice suggest that the path of free markets is preferable?

Justice and Monopoly

In order to be just, legally created rights must be grounded in brute facts, or natural law if you prefer, and not contradict other, grounded rights.[1]

As I have argued above and elsewhere, property rights in land and movables are grounded in brute facts of possession, and attempts by law to undo grounded property rights are unjust. The Soviet Union's attempts to make property 'illegal' were thus immoral. It may be similarly immoral then for states to attempt to prevent *natural* monopolies as well. A natural monopoly, as opposed to one created and supported by the state (as in the case of IP), is the accumulation of a disproportionate market share, or exclusive (captive) market through non-state-supported means. One way in which natural monopolies are created is by enterprising entrepreneurs who anticipate a market even before it exists, and are thus first to enter the market because, in fact, they created it. Another way in which natural monopolies may evolve is simply through the market's preference for one product over another. If a good product or company becomes a monopoly through the preference by consumers for that product or company, this is evidence that free markets work. Monopolistic products or companies, in a truly *free* market, are vulnerable, however, to the preferences of consumers. Assuming no barriers to entry for potential competitors, presumably better products may beat a monopoly. Of course, natural monopolies may employ other means to keep a market captive, or to keep competitors out, and some of these means may be *unjust*.

Monopolies may become unjust by violating the rights of others. No market player has a right to succeed, and many of those who argue most vehemently for IP rights claim that without such rights there is no guarantee of success even for useful products. This is true, however, for all entrepreneurial activity: there is never a guarantee of success, even for useful products. Restaurants, bakeries, newspapers, hair salons, etc., all are subject to the vagaries and whimsy of the marketplace, and they come and go with market demand and individual or group preferences. Choosing to embark on any entrepreneurial activity is inherently risky, and the best successes come with some form of innovation by which a new product or service distinguishes itself from the crowd. All of these sorts of businesses typically also require capital investments in both structures and infrastructures, and for most of these enterprises, there is no specially created, ungrounded property right to help spur their attempts to enter the market. The rights involved in free markets are the same as those that exist between human beings in other situations: to be free to act and speak within the limits of the liberty principle. There is no right to a monopoly, much less one granted or protected by a sovereign, but there is a right to *challenge* a monopoly with some competing product or service. That right ought to be free from interference. The right to enter a market exists; the right to succeed, or be granted some form of exclusivity, is utterly artificial, as well as an inefficient guarantee of market success. In fact, most innovations fail to ever succeed commercially. The Patent and Trademark Office's own studies have shown that only about 1 to 2 percent of patents ever earn their costs of filing back. Once filed, moreover, those patents sit as roadblocks to commercialization of similar, perhaps even better products and processes. This is a deprivation of grounded rights to those who would push ahead with some improved technology but for the rent demanded of the earlier, less innovative, or simply worse products. This is what happened, in fact, with Watt's steam engine. His patents were used to block better, successive versions of steam power from reaching the market until after Watt's patents expired.[2] Had Watt's monopoly on steam power (during the period of which the rate of increase of steam power production went more slowly than following the expiration of the patent) been a *natural* monopoly, earned because of the superiority of his product, and the market's economic reward for that superiority, no rights would have been violated. Moreover, innovation and economic activity would have been spurred rather than hindered. But Watt's monopoly was the result of political forces, not economic.

There is a certain irony in the attempt by governments now to curtail monopolies that have formed largely as a result of the artificial, state-sponsored system of patents and copyrights, particularly in the realm of software. The US attorney general's attempts to rein in Microsoft's monopoly over PC desktops, and claims about unfair trade practices due to 'bundling' browsers with the rest of a fully functional operating system, underline the hypocrisy of granting a patent on the one hand, and claiming to want to thwart monopolies on the other. The monopolies extended to Microsoft's products through extensive patent portfolios are largely what enabled the company to dominate the operating system market in the first place. Unfortunately, in the case of these state-supported, super monopolies, built on patent portfolios as well as potentially unfair trade practices, there is simply no way to distinguish which part of the corporate structure is built on legitimate, or natural, monopoly (due to superior products or customer preferences) versus those that are now the sole result of artificial monopolies. Defining the justice of the whole is now more or less impossible.

So what counts as just and unjust in the world of business? Some argue that these terms have no application in the Darwinistic world of capitalism. Yet, if free markets are justly founded upon the notion of autonomy, and defended against the attempts of states to interfere with natural rights to freedom of association, trade, speech, property, etc., then there are surely underlying notions of justice upon which free markets depend.

Injustice in markets occurs when the fundamental rights that demand free markets, such as individual autonomy, rights to free association, speech, and property, are disrupted. Monopolies that arise because of injustice are themselves unjust monopolies, and other market players ought to treat them accordingly, devising market mechanisms to thwart them, or imploring the intervention of states if necessary. The Microsoft irony noted above, however, illustrates that sometimes these injustices may be caused by states, particularly through their IP policies. As we have seen above, IP, inasmuch as it interferes with fundamental rights, is itself arguably an unjust mechanism by which a monopoly is not only formed, but also enabled by states to interfere with the free congress of other market players ... consumers and producers alike. While monopolies are not *per se* unjust, those that arise through curtailing the rights of others, such as by IP regimes, are not only unjust, but also likely inefficient.

There is no theoretical reason why a natural monopoly must be inefficient. In fact, monopolies may arise because they promote efficiencies in markets. When they arise through free market mechanisms, as the result of consumer demands, supply issues, and not through market manipulations by states or other forces, they may well be efficient responses to market conditions. But because these market conditions can never effectively be predicted by any market player or state, attempts to create monopolies in markets prior to the emergence of those conditions will always be inefficient.[3] The failure of state action to accurately or effectively predict or control markets is no longer just theoretically assumed, but empirically proven. The boom and bust cycle of international and local economies has not been effectively mitigated by attempts at state controls, and economies fluctuate wildly despite (or perhaps exacerbated by) attempts to regulate or otherwise manage them through fiscal and monetary policies. States are artificial monopolies of a sort, and they in turn support a multitude of artificial monopolies, such as militaries, police forces, courts, etc. These monopolies are maintained in the name of preserving order, peace, and justice. But as John Locke argued, and as revolutionary movements have taken note, when states fail to abide by the social contract of preserving our natural rights, and when they join in mechanisms and means that defy those rights, we have just cause to challenge them.[4]

The artificial monopoly enjoyed by IP holders should be open to question, and the system itself, which imbues in one party the ability to exclude others from enjoyment of natural rights to property, consciences, and expression, is subject to our questioning at least, and replacement at most. Even were we not to accept the libertarian notion that IP impinges upon our personal autonomy and rights to property, there are other injustices that arise from artificial monopolies, and which cannot be termed *injustices* in the same sense when resulting from the occurrence of natural monopolies. In the case of natural monopolies, it is natural scarcity and monopolistic control of resources that often perpetuate and grow some monopolistic practice. This sort of monopoly can become unjust when, as we mentioned above, unjust means (such as force, threats of force, theft, curtailing of speech or autonomy) are used to build or perpetuate control over a scarce resource. It goes without saying that using the state to bolster or legitimize any of these means is also unjust, and implicates the state in that injustice. Natural monopolies may act unjustly, and indeed there is nothing *per se* unnatural in any of these unjust practices. States may legitimately interfere to stop the unjust practices of natural monopolies, but they may not justly support such activities, nor may they justly deprive justly formed and maintained natural monopolies of their property. Natural monopolies must refrain from interferences that would be unjust according to our standard notions of just behaviors as elucidated already, and consistent with Mill's harm principle.

In sum, in order to be just, natural monopolies may only form through market demands, and not through manipulations that involve deprivations of natural rights. This is what also makes natural monopolies vulnerable in ways that artificial monopolies are not. Natural monopolies may only justly endure if they are founded both upon market demand and some natural scarcity (for artificial scarcities are often created through unjust means such as deceit, thefts, intimidation, etc.). If there is no scarcity at all, or if some natural scarcity is resolved and disappears (either naturally, or through productive means), then a natural monopoly will likely similarly dissolve. In a free marketplace, where natural monopolies are vulnerable to the end of scarcities, or free alterations of market demands or preferences, monopolies may be challenged by innovative new market players, natural phenomena, or customers' whims. None of these possibilities involves any elements of justice or injustice. Justice is implicated, however, as we have begun to explore above, through the appropriation and monopolization of those things that cannot justly be claimed to be property, or access to which by others is founded upon natural rights.

Ideas as Commons and Truly Free Markets

We now take for granted the background assumption that the application of ideas through processes and artifacts can be monopolized through IP. We are faced, however, with a conundrum if we believe that innovation must be promoted through state-sanctioned monopolies: if the state grants a monopoly, and it undermines or contradicts natural rights, then we must make a choice. The choice is – do we prefer to maintain the state-sanctioned monopolies and violations of natural rights to the alternative based upon utilitarian grounds, or are there deontological reasons (reasons founded upon inviolable duties) to reject the IP system because natural rights trump any utilitarian calculus? Those who argue for adopting the former choice insist that the current state and recent rate of innovation in the West is the direct consequence of IP rights, and argue that the benefits outweigh the costs. Missing from this argument, however, is any firm empirical support for the causal claim, even if we accept that the utilitarian calculus outweighs deontological concerns.

The problem is a basic one in historical studies. Although there is a temporal correlation, there is no firm evidence of causation. The past 200 years of rapid technological change do indeed seem to overlap with a period of increasing IP rights. But this overlap does not prove that IP causes development. Moreover, there are historical anomalies that suggest that innovation and development are products more of scientific advance, increased trade, and other ongoing phenomena than of laws that reward inventors with monopolies. Take, for instance, the advent of the steam age, which a recent book by William Rosen credits with the race for patents.[5] As Boldrin and Levine point out in the first few pages of *Against Intellectual Monopoly*, Watt's patents, and his tenacious legal defense of his imperfect implementations, prevented the full force of steam's potential from being realized until after Watt's patents lapsed. Patents, in the case of steam's full development, *arguably hindered* innovation. Dozens of similar examples abound, and the rapid development now of open source alternatives, embraced in many cases by corporations as well as individuals, seems to illustrate awareness of the potentially stifling nature of IP on innovation.

There have also been periods of rapid economic and technological innovation where there was no IP regime that could have been its immediate cause. Specifically, in the mid-nineteenth century, a number of European nations explicitly rejected the idea of patents, and abolished their IP laws for significant periods of time. Citing the theories of Austrian economics like Carl Menger, specifically regarding the nature of value, many economists in mid-nineteenth century Europe embraced a radical conception of free markets. Patents and copyrights were considered to be inefficient given that they do not allow a free market to function for their subject matters, but rather create artificial scarcity, and monopolistic prices that have nothing to do with individual valuation, which is at the heart of Austrian value theory. From about 1869 to 1912, The Netherlands had no patent law at all, while at various other times from about the mid-1800s to the turn of the century, the United Kingdom, Switzerland, and Germany all shirked patents altogether, or in various specific realms of innovation.[6] There is evidence, moreover, that during the patent-free periods of these European economies, innovation and economic growth proceeded apace.[7] Recent evidence has also suggested that patents may actually hinder innovation in certain areas, which was the thesis of many anti-patent advocates during the mid-nineteenth century as well as today.[8] This evidence, and the lack of evidence of causation where there is only bare correlation, suggests that those who argue that patents are efficient or necessary bear a heavy burden, especially in light of their implications for justice.

Ideas, it has been said, want to be free. Indeed, they are free. Once expressed, nothing can contain them, and positive laws enacted in order to curtail their uses defy natural rights, as we have seen above. They are not clearly efficient, nor are these laws just where they conflict with grounded rights. Rather, justice argues for their complete liberation at every turn. I have argued that ideas are part of what I call a 'commons by necessity', which must be distinguished from commons that we *create* out of things that otherwise are susceptible to just ownership. Free markets depend upon commons by necessity such as ideas, and we are beginning to see how attempts to enclose unencloseable spaces, or to create means of possession of things that cannot meaningfully be possessed, impede free markets, inhibit scientific progress, and thwart justice. By opening back up the marketplace of ideas, a truly free market, both unsupported by and unimpeded by manipulation by states, will allow innovators to reap the rewards of similarly open science, and create new inventions with less friction. The marketplace for *things* still relies upon scarcity to drive market prices, as there are always, at any one time, only so many things that can be traded and possessed, and always some limited number of producers of those things.

The Scientific Commons and the Marketplace

Much has changed in the relationship between what we once might have called *basic science* and its traditional realm of commercialization through *technology*. Science, on the one hand, has been characterized over time as the pursuit of knowledge about nature: an attempt to codify nature's laws. Technology, on the other hand, is the application of those laws through artifacts to utilitarian ends. There is, of course, much overlap between the domains of science and technology, as well as a fair gap between the two. For much of the history of scientific progress, there was a relatively clear social disjunction between the methods and manners of science, and those of industry. Even while innovation and invention were informed, and in a sense driven, by scientific discovery, science was typically pursued in academia, impelled largely by perception of the nobility of scientific discovery and academic standing, while technology was developed by entrepreneurs for markets and money. Although scientists have always been subject to the same human weaknesses we all are susceptible to, secret-keeping and monopoly over ideas have always been anathema to the long-term pursuit of science as well as scientific prowess. No discovery can be lauded for long if it is revealed to be false, and sharing data is vitally necessary to confirm the truth of a discovery. The currency of scientific progress is confirmability and stature in the community, which comes with successful unique and foundational discoveries about nature's truths.

Basic science requires openness at some point. While many who promote the necessity for patents call attention to the role that patents play in requiring disclosure, and thus informing the public about the technique employed in any invention, basic science has always thrived upon disclosure, even while technology has often thrived on secrecy. Indeed, if there were a danger that scientific discovery would by its nature *exclude* those other than discoverers from knowledge of nature's truths, then there would be an excellent argument that justice would favor patents in order to promote disclosure. Surely it is a human right to know the discovered truth about the nature of the universe and its laws. But there is no more evidence for this notion than there is for the notion that patents are efficient or necessary for innovation. Scientific method depends upon disclosure, and scientists who do not disclose will not be trusted among their peers, nor their allegations of discovery heeded, nor their stature elevated. Ever since the Royal Society and similar academies of science began, so did their publications and meetings begin as a means for the dissemination of methods and results so that peers could judge both, and hopefully replicate experiments or prove them to be false.

Science, as an open institution, has undergone some troubling changes in the last thirty years, especially in the United States. The Bayh–Dole Act in the United States was passed in order to encourage universities to speed the transfer of knowledge to companies in order both to spur the economy and to provide an alternate revenue stream for universities. The notion of 'technology transfer', which was supposed to help universities identify potentially profitable scientific work, and move it toward productive technologies, was instituted by the Bayh–Dole Act in 1980.[9] Among its innovations was the new ability for universities and their scientists to reap direct profits from inventions, even where the basic science might have been conducted with taxpayer funds through federal research grants (as most science is funded, even now). The Bayh–Dole Act fundamentally altered the social and economic climate surrounding basic research, and is being replicated throughout the world as national governments seek to provide alternate sources of funds for universities and research. But even before this turn, funding for science had changed dramatically since the Second World War as opposed to the centuries of academic science before it. Modern 'big' science, on the model of the success of the Manhattan Project, and as evangelized by Vannavar Bush following the Second World War, built large, government-sponsored funding bureaucracies into the existing academic model, and yielded many of the breakthroughs, both scientific and technical, that led the United States economically in the post-war era.

But governmentally funded science has proven to be a double-edged sword, as academic scientists once propelled by their perceived duties to scientific discovery itself, as well as their own statures among their peers as well as careers, began to chase ever-expanding grants for funding. Grant funding became as important, or more important for universities, as publication.

Grants supported swelling science and technical departments, and became a beacon and symbol of universities and departments that excelled. Government grants also often come with a catch. Basic science was no longer a sufficient impetus for research, and general, foundational knowledge is hard to justify to taxpayers, even while the aesthetics of knowledge for knowledge's sake is compelling enough to drive many into scientific careers in the first place. Funding, moreover, does not guarantee success, or even value. Much of science is comprised of dead ends, hypotheses that turn out to be wrong, and sometimes even delusion. The race for grants and notoriety has led a number of prominent scientists to public disgrace, including Martin Fleischmann of the University of Southampton and Stanley Pons of the University of Utah, who, in 1989, prematurely announced achieving cold fusion in an attempt to scoop other researchers who were viewing similar results as part of similar research programs. Their announcement was hoped by the University to be able to help win priority for potential patents as well as research grants, and the University encouraged the scientists to announce their (apparent) discovery via press conference, rather than through the age-old method of peer-reviewed publication and public scrutiny. When laboratories throughout the world failed to be able to replicate their results, Pons and Fleischmann were publicly disgraced, although they continue to research the effect they witnessed. Similarly 'pathological' science has occurred before the age of giant government grants and patents, but the added pressure and enticement of these grants, as well as their importance to academic institutions, continue to threaten to pervert the ideal methods of scientific discovery. The race for grant funding exacerbates whatever human frailties exist that otherwise thwart openness, and every instance of failures to pursue science without secret, obfuscation, or other opacity (whatever the motivation) has proven in many ways to be detrimental to scientific discovery, or at least to the institutions of science.[10]

A lack of openness makes sense in commerce (sometimes), and trade secrets and other means of obfuscation or deception are to be expected in the race for customers. But creating technologies and discovering scientific truths are still epistemologically separate domains, even where our institutions now force those lines to become blurred. Openness 'upstream' closer to the point of discovery will only tend to lubricate invention 'downstream', where innovators can apply well-understood and confirmed scientific principles to new techniques and inventions. This is how science and technology have interacted for hundreds of years, before the various perversions of the modern era tried to flatten the stream, and this age-old division of science and technology is *just*. Scientific truths belong in the public domain, facts about the workings of nature cannot be contained, cannot be monopolized, cannot justly be owned. Rather, if anything is to be granted a government-backed monopoly (though the necessity of such monopolies has yet to be demonstrated) it must be limited to those things that are clearly human creations, and not natural facts, laws, processes, or products. Science belongs in the public domain, and functions best when it is conducted there separate and apart from industry. Industry can thrive if science proceeds apace, and can reap its rewards on top of scientific discoveries, maintaining where possible the separateness of their mildly overlapping magisteria.

One might object that science, as it is conducted today, simply cannot thrive without funding of the sort to which it has been accustomed. This I truly doubt. Many large research projects currently funded by large government grants may well be funded by private philanthropy, as science was more or less funded before the current big science/big government paradigm. Is it not possible too that faced with the trough of public funding through large grants, research budgets have expanded to fit the funding, rather than according to some natural need? The merger of academia with grants and now with business has, instead, produced a hybrid economy of science, with competing directions from each party, changing the direction of scientific pursuits, and altering the intentions of researchers. The goal is now often the next big grant, and impelled by this goal, the methods of inquiry and communication of science to both the public and funders have been perverted to salesmanship at its worst, and even at its best, imbued with the tactics of marketing.

Scientific inquiry ought not to be torn between the present conflicts that have morphed it from its prior forms, and decreased public confidence in its methods and practitioners. Neither science nor industry needs the support of big government, either through funding agencies or through IP laws. The pursuit of natural truths is impetus enough, or should be, for the best scientists to delve into the mysteries of the universe. Academic stature sufficed for generations of scientists, and Albert Einstein's greatest theories were dreamed of without research grants. Even the first experiments that began to confirm relativity were funded largely through philanthropy, scientific institutes, and university budgets sponsoring their own scientists' research. By altering the nature of scientific inquiry, and generating a climate in which conflicts of interest abound, governments and academic institutions have acted unjustly, selling out now the public domain of science to the highest bidders in so-called 'public, private partnerships'. These monsters of tech-transfer are essentially double-dip subsidies for business from governments, and the promise of riches dangled in front of researchers is used as a carrot to entice their cooperation, all in the hopes of enriching university endowments, but at the expense of basic science. First, the science is publicly funded through government research grants, at taxpayer expense. Next, the market is skewed through patents that create artificial scarcity and enable the setting of monopolistic prices. Now, at the double expense of the public, scientific knowledge has been moved out of the public domain, as indeed publishing results takes a back seat to building a workable business model (often by delaying publishing), and the public further subsidizes some hybrid private enterprise both through the public funding of the 'basic' science and the private appropriation of that knowledge for private ends through the institutions and mechanisms of the state. This appropriation is unjust, and argues for a return to the basic formulation of science and industry that sufficed before the Second World War, and that thrives still among scientists who work either without the benefit (or shackles) of public funding through large government grants, supported perhaps by their academic institutions, philanthropy, or both.

Justice and Law: Rejecting Positivism

Law was not always so divorced from justice. There has been a trend first in legal philosophy, now adopted more or less as the primary model of a legal education, toward what is called *legal positivism*. In the nineteenth and early twentieth centuries, the natural foundations of law began to be challenged, especially in Anglo-Saxon legal scholarship. For more than a thousand years, legal rules were devised by sovereigns who were themselves held to be imbued with moral authority for rulemaking by deities. Of course, with the spread of the Enlightenment, and the fall of various sovereigns at the hands of liberal revolutions, the basis for valid, moral rulemaking and enforcement began to shift. With Locke, Hobbes, Rousseau, and other modern liberal political theorists came a new vision of the basis for natural law, one that extended natural law theory beyond the simplistic, sovereign-based dogma of old, to a more consistent set of tenets. Natural law, it was argued, grounded the validity of legal rules in duties and obligations dictated by the fabric of nature (whether or not one accepted some predicate deity), and even sovereigns were subject to the dictates of nature's laws. This shift in thinking reflected a shift in scientific and theological thought, roughly reflecting the move from an involved, acting creator to a distant, detached, clock-maker creator, who sets the world in motion and then steps back. The foundations of modern liberalism included a notion of natural law. The revolutions sparked by Locke *et al.*, were legitimized by the violation of subjects' naturally endowed rights to life, liberty, and property, and these natural rights formed the basis for modern liberal democracies, both in their constitutions and in their laws. In a world in which just law derives from nature, there is a solid connection between law and morality. But a new trend emerged in the late Enlightenment, when philosophers and political theorists began to question the foundations of just law, as well as to reformulate approaches to ethics and justice themselves.

The British philosopher Jeremy Bentham led the move away from natural law theory in arguing against deontology (which accepts the existence of categorical ethical duties) as an ethical foundation. Bentham, seeking to make more scientific the study of ethics, rejected nature as a foundation of duties, and formulated the modern ethical approach we call Utilitarianism. He is well known for calling natural law theory 'nonsense upon stilts'. His objections to natural law were epistemological, as he argued we can never rightly suppose we know the intentions of a creator, nor can duty-based theories of ethics like those of Immanuel Kant ever trace back sufficiently enough to provide a solid justification for accepting any particular duty *a priori*. Indeed, a weakness of deontology is the leap from presupposing the existence of a certain duty, and reconciling its existence with contradictory duties, or converse duties that appear to arise in exceptional situations. Utilitarianism does not presuppose the existence of categorical duties, but rather argues that the ethical compulsion lies not in intent or duty, but rather in consequences. The epistemological argument is clear: consequences can be more or less predicted, and measured *post hoc*. Intentions can never be similarly measured. In the scientific vein of the time, Bentham sought to make the pursuit of legal and ethical theory measurable, and viewed a solid, measurable basis for judging an action only in its consequences, namely the amount of net happiness produced. For Bentham, the good can be judged based solely upon whether it increases net happiness, and duties, intentions, or other epistemologically unapproachable matters need not be consulted. The implications for legal rulemaking are obvious, and in the absence of a natural foundation for just laws, legal theorists began to reimagine the role and scope of legal theory.

According to John Austin, the validity of legal rules can only be judged according to their proper foundation of their enactment by a sovereign. A sovereign is one who is recognized as such by a majority, and *just* laws are merely the sovereign's valid enactments, backed by sanctions. No further inquiry of judgment as to the content of legal rules can be made as there is no extraneous basis by which rules can be judged to be right or wrong, morally speaking. Rulemaking is valid so long as the sovereign is the majority-recognized sovereign, and has no higher sovereign, and so long as the rules are backed by the promise of sanctions in case of their violation.[11] Even while a continental positivism of a sort was being formulated by Hans Kelsen, in which at least some solid basis for recognizing a valid sovereign is posited (a '*grundnorm*'),[12] the Anglo-Saxon school of positivism became solidified with the work of H. L. A. Hart.

Hart nicely categorizes types of rules, distinguishing among primary rules (which direct action) and secondary rules (which address procedures). But in direct opposition to Kelsen, Hart rejects the theory of a *grundnorm*, and does nothing to resolve what seems now to be a significant gap in positive legal theory: reconciling rules with a notion of justice. As opposed to the neo-Kantian approach of the twentieth century's most prominent legal scholar outside of the positivist tradition, John Rawls, legal positivists do not see inquiring into the just foundations of legal rules, outside of the valid enactments of sovereigns according to established procedures, as being a coherent area of inquiry.[13]

Legal positivism is the political extension of the ethical theory of utilitarianism. With legal positivism, we need not concern ourselves with metaphysical questions of right or wrong, just or unjust, but can focus instead on epistemologically approachable questions regarding the results of our actions, and whether they accord with our preferences. Legal positivism is the dominant theoretical paradigm in Anglo-Saxon law schools, and it is bolstered by various trends in politics, including concerns with pluralism and multi-culturalism. Natural law theory is vulnerable to critique where various cultural, religious, ethnic, or philosophical backgrounds confront problems from differing viewpoints. Adopting the natural law justification for a rule that contradicts some religious, ethnic, cultural, or philosophical viewpoint means arguing for the error of someone's point of view. But as we live in increasingly pluralistic societies, with ever more multicultural populations, asserting some particular paradigm to be correct risks eroding what many conceive to be a foundation of liberalism: the freedom of conscience. A basic tenet of our modern liberal democracies is necessarily that people are entitled to their opinions, points of view, and to express their beliefs. Thus, states ought not then to criticize the foundations of those beliefs, or force citizens to ascribe to a particular point of view. Because positive legal theory embraces the notion that a law is valid so long as it is enacted by a sovereign and backed by sanctions, then there is no further basis to question the validity of a validly enacted rule. The freedom of conscience of those who either support or defy the rule is preserved, because no judgment about the underlying justice of that rule may be made. We can only classify people as rule-followers or rule-breakers, not as just or unjust, and the basis of valid rules need not be traced to any natural, immutable principle. Pluralism and multiculturalism are preserved both within nations and among them, as ethics and rules are completely divorced. A rule-breaker cannot be judged to be immoral, and rules we do not like can be changed without reflection upon metaphysical issues of justice or the good. Lawmaking can be scientifically accomplished by looking at a list of projected consequences, and applying those rules that maximize the consequences we prefer.

Legal positivism is vulnerable to attacks based upon history, and our cultural, national, and international reactions to perceived injustices within sovereign states, as well as among them. These same attacks are consistent with criticisms of utilitarian ethical theory. Namely, if we are only guided by the consequences as a guide to action, then on what moral basis must minorities be protected? In utilitarianism, as in legal positivism, there is no theoretical basis to necessitate the protection of a minority. In classical utilitarianism, the right thing to do is that which increases happiness (maximizes utility) overall. Positive legal theorists similarly must recognize the validity of an enactment if it is enacted by a valid sovereign (supported by a majority) and backed up with sanctions. Countless examples of potential injustices can be named, both historical and hypothetical.

A Vacuum of Justice

As with utilitarianism, or moral relativism, positive legal theory leaves open the difficult problem of determining when a particular action, or intention, is morally wrong. In fact, in none of these theories is the notion of 'moral wrongness' even comprehensible. Things may or may not be acceptable in specific contexts, or may be valued for their effect on general utility (inasmuch as it might be measured or measurable), but notions of right or wrong, as the terms are traditionally used in ethical theory, are not *per se* applicable. Although students of ethics are taught about utilitarianism, and ethics scholars, or applied ethicists, must resort at times to the hedonic calculus in resolving ethical dilemmas, the end result of such a calculus will always be some determination about what one should do in order to increase general utility (happiness), and not clearly an ethical judgment about what is right or good in a moral sense. This is because each decision is necessarily contingent, and hypothetical, as opposed to decisions made according to deontological theory, which are categorical and apply to every such action or intention. Some of the weaknesses of utilitarianism in creating systems of justice are noted by John Rawls, and other modern Kantian or neo-Kantian scholars of law and ethics. These weaknesses make it difficult to argue that positive legal theory, or utilitarianism, can lead a society to a state fairly called *just*.[14] The term justice implies some accord with notions of morality. In modern constitutional parlance, there are two forms or aspects of justice: substantive and procedural. Procedural justice means simply that for every person who becomes involved in a criminal or civil judicial matter, the procedures employed are employed equally, and fairly, and their content is transparent, and purposes clear. Substantive justice is more complex, and the notion implies some accord with some higher law. If a law fails to fulfill the requirements of substantive justice, it may justly be struck down. Substantive justice is a measure by which both constitutions and legislation may be judged, and according to which they may fail.

Given the weaknesses of positive legal theory in providing a solid context in which justice can be evaluated, or by which just legal systems and their rules can be imposed, why does it continue to thrive in legal scholarship and political theory? One explanation may be that legal and political scholars have abandoned the quaint, Kantian notions of categorical right and wrong, and have embraced a utilitarian world view. It seems to be that in so doing, and in simultaneously accepting the Rawlsian notion of distributive justice (as indeed some of these same scholars and theorists seem to do), they are trapped in a contradiction. Rawlsian distributive justice depends upon accepting the notion of categorical duties, including the duty to treat everyone as an end, and not merely as a means to an end. Another categorical duty under Rawls is to treat everyone with equal dignity. But Rawls accepts more or less the Kantian explanation for the existence of these duties, arguing that we would arrive at these duties in forming a society if we place ourselves in the 'original position' behind his hypothetical 'veil of ignorance' from which vantage point we have no idea of who we might be in a society. Kant's categorical imperative is arrived at by a different heuristic, but the content is the same: we have to be able to successfully universalize an imperative without contradiction in order for a rule to be moral. Neither Rawls nor Kant judges the morality of a rule according to consequences, and Rawls is thus generally agreed to be a neo-Kantian, as he himself at times contends.

If justice becomes hypothetical and contingent, as it must under a utilitarian/positivist perspective, then rulemaking will be similarly contingent, and may even fail to be just. Just as Bentham insisted, the link between lawmaking and morality must be completely severed, and decisions about the justice of rulemaking or rulemakers must be limited to procedural matters. Arguably, no coherent system of substantive justice could be based upon utility as a measure or standard by which just laws could be created. The barriers are epistemological (the calculus cannot be carried out to sufficient exactness, either over and across populations, or through time), as well as ontological: the calculus does not tell us what is good or right, but merely what we should do in a certain situation to maximize happiness. One blatant gap in accepting the logic of the latter statement being somehow a coherent foundation for a just system is that it relies upon a categorical rule, one which cannot be adjudged scientifically, namely that happiness is a sound basis for moral decision-making. This itself implies a categorical rather than hypothetical grounding which must be taken as an axiom. Because of this and similar logical gaps in utilitarianism, and unacceptable practical consequences of accepting a pure utilitarian basis for ethical decision-making, legal positivism stands on similarly shaky ground. The fact is that neither rulemakers nor ordinary persons function as though there is no greater grounding for just rules than utility. There are clear, historical instances both within and among nations where actors (both individuals and states) have done things that are clearly unjust, but which they justified to themselves as warranted based upon their perceived effects upon general happiness. Evolutionary psychology may hold the key as to why we consider certain intentional states and actions to be wrong *per se*, but the fact of this acceptance is recognized in constitutions and in courts. It is the impetus behind the slow march toward greater freedom, and more perfect systems of justice. The general recognition that, despite the arguments of legal positivists, there are certain categorically wrong actions and intentions is what has enabled constitutional change as well as liberal revolutions, and it is what has made these historical moments good.

Progress and Justice: Embracing a Natural Basis for the Good

Even while the Austrian Kelsen was making a case for positivism with some *grundnorm*, another little-known Austrian lawyer-philosopher was arguing for a grounding of valid legal norms in natural states of affairs. Adolf Reinach's 'The Apriori Foundations of the Civil Law' explains that valid legal rules are logically dependent upon some natural state of affairs, which he calls 'grounding'. He argues that the law of contracts, for instance, is logically necessary because it is grounded in the simple facts about the genesis of duties out of promises. Prior to lawmaking, the acts and intentions surrounding the human activity of promising generate claims and obligations. These claims and obligations disappear upon the fulfillment of the promise. Contract law is thus grounded in natural phenomena, and this he equates with the sort of logical necessity that makes the facts of mathematics true. No just enactment, he argues, could invalidate the claims and obligations that naturally arise from promising, just as no valid enactment could make 2 + 2 equal 5.[15]

Reinach and others have since extended the notion of grounding to other types of law, including property law. Some of us have argued that our rights to ownership of property arise from the brute facts of possession with the same sort of logical necessity by which duties and claims arise from promises. This same argument is extended by Austrian philosophers of the same vein over rights to autonomy, which is rooted in rights of 'self-possession'. There is a revival of sorts, for Austrian philosophy of law, and other similar schools of thought that oppose positivism. Ronald Dworkin stands as an example in legal jurisprudence.[16] Natural law is not dead, it has been naturalized. No longer is it dependent upon any particular ideology, theology, or philosophy.

Natural law is the basis for many of the claims above: that property in tokens is grounded and claims to ownership of tokens are just; that claims to IP ownership are not grounded and conflict with grounded claims; and that appropriation of knowledge and ideas in the public domain, aided by the state, is unjust. These arguments above depend upon a view of natural law grounded in brute facts, but utilitarian arguments might well also make the same case. There are pragmatic reasons why, especially in the age of converging technologies, the old models that once propelled scientific discovery, and allowed private, commercial application of scientific discoveries through the development (without state support by grants or patents) of new and useful technologies for the marketplace, might now be preferable to the state-capitalist methods now embraced.

Rights to Tokens and Exchanges of Types: Pragmatic and Theoretical Approaches to Markets without Scarcity

Is there a way, then, to reconcile our demands for justice with concerns about sufficient motivation and institutional support for scientific and technical progress? Can we respect property rights, promote investment in both science and innovation, and develop markets without subsidies? It has been approached before. Totally free markets have never truly existed, as regulations both within and among states have always modified markets, and intruded upon their free functioning. Nonetheless, we can envision a marketplace in which science is pursued through private investment, academic institutional support, and philanthropy, where knowledge of scientific discoveries flows freely into the public domain, and where commercial uses of that knowledge are made through developing new products. The incentive for industry is to capture a market share quickly with a product people want. Innovation is a special *service*, performed by entrepreneurs and inventors who envision some new way to fulfill a market need. That service should be rewarded commensurate with the market's appreciation of the usefulness of new products in fulfilling actual needs or desires. In fact, that is how most commerce proceeds. If a product enters the marketplace, and is perceived as valuable by consumers, then the scarcity of the product and the desire of consumers to purchase it help to drive the pricing of the product in the market. The same is true for services.

Most services are unpatentable and unpatented (although patent attorneys are trying hard to turn services into patentable subject matter). Yet services account for a growing portion of GDP throughout both the industrialized West and developing economies. Many web and software applications are just as much, if not more, service than product. Companies that purchase even open source software products also purchase some service package from a vendor, and it is typically in services that money is made for open source products. Even where products will doubtless suffer less from the effects of scarcity in a future of nanowares, given localized production, and eventually true MNT, the service of *designing* and *creating* useful new products, and of turning science into technology, will continue to be scarce. This scarcity will properly drive market pricing, combined with market demand for the results of innovation. What is needed is not a new form of protection for those who will be creating new products, but a rational means of *understanding* the nature of future transactions in types and tokens.

When you receive a service, do you pay for it? Who sets the value, and how does the one providing the service know that you will pay for it upon completion? Unlike the exchange of a product, which is generally initiated by payment, and concludes with delivery of the product, payment for services usually coincides with the service's completion. Without more to guarantee it, only *trust* and a sense of *obligation* ensure that the service provider will be paid properly. In more formal exchanges, services are secured by contracts. In fact, treating the act of creation more like provision of a service is already one method by which creators of artistic works are profiting, even without the direct exchange of products. Some examples include:[17]

- In July 2009, Erin McKeown, a folk-rock singer-songwriter, launched a Live Web Concert Series, entitled 'Cabin Fever', to raise money for her new album 'Hundreds of Lions'. Erin produced and broadcast live over the internet four shows – one from her living room, one from her front porch, one from her front yard, and even one right from the middle of a river. Viewers could buy a 'ticket' online that would allow them to stream the concert. Erin also included a 'donate' button in case fans wanted to contribute more to her album.

- Erin profited directly from these online videos, but other artists have profited indirectly by using viral videos as marketing tools. Perhaps the most well-known viral video success is OK Go. Formed in 1998, the band had achieved some critical acclaim in the United Kingdom, but was relatively unknown in America. In 2005, however, they hit it big when their music video for their single 'Here It Goes Again' was uploaded onto YouTube and instantly became a classic internet meme. The video, featuring the band members performing a synchronized dance on moving treadmills, led to a huge spike in CD sales, sold-out shows, a Grammy Award for Best Short Form Music Video, and instant fame.[18] The band considers the video so integral to its initial success that when its record label disabled the embedding feature in their most recent YouTube video, they launched a campaign begging their label to reconsider. After massive media coverage, complete with a New York Times op-ed by Damien Kulash, OK Go's lead singer,[19] the label renewed the embedding feature and allowed the video to be endlessly reposted on blogs and websites across the internet.[20]

- Another viral video success story is Tally Hall. One of the members of the band was a video and graphic design major at the University of Michigan; for a class project, he had created a music video for the band's song 'Banana Man' on a website called http://www.AlbinoBlackSheep.com. After that, the band's website started getting thousands of hits per month, even though the band had never played a show outside of Ann Arbor, their hometown. A bio written about Tally Hall for SXSW 2007 includes the band talking about the impact of their videos: 'We didn't really realize what was going on outside of Ann Arbor', says Ross [of Tally Hall]. 'Then our website started crashing each month from the bandwidth and it became obvious that there were people outside of Ann Arbor that were listening to the music.'[21] The band's fame began on the internet, and they stayed true to that medium as a marketing tool. They were able to keep their audience interested by releasing dozens of small, internet-only mini-videos on their website, maintaining a livejournal blog, and creating a discussion board. They even launched an internet show, called T.H.I.S., which will air biweekly and last ten episodes. The promotion they received from the short videos catapulted them into the national arena and, in combination with heavy touring, led to a major label deal.

Tiered pricing models

- Artists have also realized that the one-size-fits-all pricing model of the traditional content industry is not the most efficient compensation model for a niche-market economy. Trent Reznor of Nine Inch Nails, after ending his record label contract, decided to release *Ghosts I–IV* on his website without any label or retail assistance. The music was available at five different price tiers: for $5.00, you got to download 'all 36 tracks in a variety of digital formats, including a 40 page PDF';

for $10, you got two physical CDs and a sixteen-page audio booklet; for $75, you got two CDs, two DVDs containing wav files and stereo mixes (allowing the user to reconstruct and remix the album), and a forty-eight-page book of photography; for $300 you got a limited edition box set personally signed by Trent Reznor and containing two CDs, two DVDs, a book of photography, and two exclusive art prints from the album.[22] The fifth pricing tier allowed fans to simply download DRM-free MP3s for free.[23] The album ended up earning $1.6 million dollars from 781,917 orders.[24]

- Jill Sobule, a singer-songwriter who enjoyed modest popularity during the mid-1990s, also successfully monetized her art using a similar tiered pricing model. In 2008, Jill raised money for her record, 'California Years', by giving fans twelve different contribution options ranging from $10 to $10,000.[25] For example, fans would receive a digital download of the album when released for $10. For $200, one could receive free admission to all of her shows in 2008. She would write someone a theme song for $1,000, and she would let someone sing on the album for $10,000. Jill ended up raising $89,000 for the record.[26]

Each of these examples above represents a significant shift in perspective of the role of creator in the marketplace, and the nature of his or her creation as a commodity. Faced with the challenges posed by peer-to-peer distribution of music over the internet, and falling prospects of large record companies and their contracts with small bands, artists have become entrepreneurs, devising new methods of reaching audiences and getting paid for their creations or even for their services. These examples should be viewed as encouraging models for future exchanges of nanowares, or the types that comprise them, by viewing the creator's provision of a *service* as valuable, even while no law can justly contrive IP rights over the types inherent in the tokens. We should keep these examples in mind in subsequent chapters as we examine whether and how we might encourage innovation, ensure free markets, and develop future marketplaces for the exchange of goods in the nanowares future. But before we conclude with a fully formed utopian vision of how such a marketplace will work, or what it might look like, we must finally confront the potential for dystopia or destruction posed by converging technologies, and consider whether and to what extent states and regulations or ethical duties properly conceived might prevent disaster.

Case Study

Development and Innovation: A New Approach?

Peruvian economist Hernando de Soto's work has brought millions of impoverished people into a worldwide economy by focusing on another, more ancient type of property: land. His idea is that people want to own that which they build on and improve, and that if economic systems don't recognize that occupation and improvement of land creates rights to ownership which should be legally recognized, then underground economies prevent the full development of capital markets, and ensure that chronically impoverished people will remain so.

De Soto's successful implementations of the Lockean theory that occupation and improvement gives previously unrealized value above and beyond often Byzantine methods of land registration and tenancy have bridged a gap between theory and practice.

This innovation is succeeding in developing nations to legitimize previously underground economies and bring their players into the mainstream capital markets. The United Nations recently appointed de Soto and Madeleine Albright to head the new 'High Level Commission on Legal Empowerment and The Poor', with the strong support of Kofi Annan, who stated that de Soto 'is absolutely right, that we need to rethink how we capture economic growth and development'. But what of those who lack even occupation, improvement, or labor on land? Or what of those whose improvements and innovation are less tangible than a plot of corn, or a successful stall for selling handcrafted goods? What of the impoverished, developing world's Thomas Edison who lacks any capital but has good ideas – the domain of intellectual property regimes (IPR)?

In most of the developed world, capital has changed dramatically in the last few decades. While tangible assets are valuable and companies may borrow against them, the most valuable assets in developed markets are typically ideas themselves. Intellectual capital, broadly defined, is the stock of assets often recognized by copyrights, trade secrets, patents, and trademarks. The value of most modern corporations is largely tied up in intangible assets which have only recently been exploited effectively (though maybe not always efficiently) by the creation of complex schemes of IP protection.

It seems unlikely that potentially untapped intellectual capital can be successfully realized only by imposing Western systems of IP protection in the poorest nations on earth. US and European style IPR may not be efficient, requiring complicated searching and registration, or even expensive litigation. Untapped intellectual capital in the developing world must be treated like that hidden value which de Soto is uncovering, by building institutions sensitive to local needs and culture. No one yet has applied de Soto's insights and success to the realm of intellectual capital.

In the West, three major forms of IP protection have been developed: copyright, patent, and trademark. These legal schemes are radically different in form, accessibility, scope, and ease of access. All of these schemes depend now on registration systems managed by governments. Developing markets surely hold untapped intellectual capital that could be unleashed by developing workable IP regimes adapted to developing nations' particular histories, markets, customs, and economies. That capital will be freed in ways which also cooperate with rather than undermine modern western IP regimes, and which provide fresh, new reserves of wealth for even those who do not own land, but who have innovative ideas and entrepreneurial ambition.

Unlike the Lockean basis argued to justify rights to ownership based upon occupation and improvement of land, IP regimes are purely creatures of the positive law, only historically recent in origin, and emanating almost entirely from Western Europe. Ideas are by nature non-exclusive and non-rivalrous. Thus, while my occupation and improvement of a parcel grounds my rights to continued occupation, and thus to its value as capital, IP is by nature non-exclusive and non-rivalrous. Your use of my ideas requires no violence, no force, and deprives me of nothing physical. IP regimes were created to provide some measure of exclusive control to ideas, which are otherwise free to be used by anyone once they are expressed. Applying de Soto's methodology regarding land requires also some new theoretical basis for extending rights to ideas.

Case Study: *Development and Innovation: A New Approach?* (continued)

De Soto's insight is that legal and cultural institutions can sometimes hinder access to capital because of their complexity, costs, or other impediments. Western style IP regimes would seem to be likely to pose such barriers to entry given their current costs and complexity. Developing countries are not homogeneous: some are more technologically advanced, some are less so, and their IPR policies should take into account context-specific features (Commission on Intellectual Property Rights 2002: 1). The Commission on IPR highlights that there is a need to design IPR regimes and policies that take into account context-specific features of each developing country, as what works in India may not work in Brazil or Botswana (Commission on Intellectual Property Rights 2002: 1). This initiative moves the debate on IPR policies beyond the previously prevailing 'one-size-fits-all' approach and opens room for designing IPR policies in order to better meet the needs of developing countries. The suggestions advanced by the Commission on IPR are nevertheless within the realm of the existing IPR regimes and rules (Commission on Intellectual Property Rights 2002).

In 2007 the Dutch government launched the 'Millennium Agreement' on Intellectual Property Rights and Development (IICD 2007). One of the three main projects within the Millennium Agreement analyzes the extent to which the current IPR regimes work to the disadvantage of developing countries by taking the case of selected African countries (IICD 2007). The aim of the Millennium Agreement is to identify areas in which the current IPR regimes work in disadvantage of developing countries and to recommend how IPR can be modified where necessary in order to be effectively applied to the local African context with the ultimate aim to best achieve and not hinder the realization of MDGs.

De Soto's actual work in development has yielded tangible results. His engagements with developing economies have built upon and helped confirm his theory that institutional barriers often prevent access to capital. The main thesis driving de Soto's and the Institute for Liberal Democracy's work is: a thriving market cannot exist if most potential players in such a market are pushed outside by institutions. The problem he has identified with many developing economies is that complex institutional arrangements, meant to benefit an elite minority, keep many legitimate entrepreneurs from access to the products and value of their labor and ownership, keeping them out of markets, forcing them into illegitimate markets, and impeding their access to capital. By researching local conditions, laws, institutions, and customs in developing nations, de Soto has identified roadblocks that prevent the entrepreneurial poor from realizing the full value of their occupation and improvement of land. Once identified, de Soto's methodology has provided guidance to these economies in removing impediments to legal ownership of, and therefore access to, landed capital. The same general framework has not yet been applied to intellectual capital.

Just as institutional impediments, once identified, have been successfully removed or overcome, enabling access to landed capital in developing economies, so too should IP and innovation experts identify institutional impediments that may exist, blocking access to intellectual capital – profitable ideas. As with de Soto's work in reconfiguring cadastral registration regimes, adjusting IPR in developing economies cannot be approached as though 'one size fits all', and should take into account local cultural and social contexts, providing greater ability and incentive for entrepreneurs to profit from their innovative ideas. Even where western IP regimes might prove to be inadequate, local solutions that can rectify lack of access to capital markets should be configured.

Research Question:
Can adjusting institutional norms according to de Soto's insights afford greater access to intellectual capital in developing economies? Will standard IP regimes suffice for liberating untapped intellectual capital, or will technologies encompassed in the broad category 'nanowares' be sufficient? What institutional frameworks will fit best with nanowares to enable those in the developing world to have the greatest access to their intellectual capital?

CHAPTER EIGHT

Nanotech Nightmares[1]

What is the regulatory environment that will best encourage the development of nanowares? Clearly, I believe that significant changes need to be made in our IP regimes, or in how we choose to relate to existing regimes, to best achieve the full promise of the technology. But the potential for physical harms must also be fairly considered, and some balance between private and public action should be achieved. We touched briefly already on the visions and likelihood of various nightmare scenarios emerging from converging technologies. These concerns, while likely still remote, are real and implied by past trajectories of science, technology, and industry, as well as by history's noteworthy lapses in human judgment and ethics. Someday, someone will create new means to cause harms, and these harms may threaten either individuals or populations. They may even threaten the species, or all species. Our technologies, while inherently neutral, are often used for ends we neither predict nor wish, and eventually perverted to serve what might well be termed *evil* ends. We should take seriously the duties of scientists and innovators to avoid the unintended but evil potential uses of their work, as well as to confront publicly these possibilities both to educate and to prepare.

Central to the discussion of the dangerous or deadly potentials from converging technologies is the question of whether and to what extent scientists and engineers owe duties to the public at large, or to individual consumers. Moreover, do governments have duties to restrict the development and uses of certain technologies, to regulate them, or simply to provide oversight? Bill Joy's grey-goo scenario stands at one extreme of the dangers posed by true molecular nanotechnology (MNT), but there are more subtle, potentially harmful uses to which nanowares might be directed. A whole new realm of intrusion could be ushered in by miniature machines capable of monitoring us without our notice. An arms race, of the type envisioned by Neal Stephenson in *The Diamond Age*, where nations, corporations, or even individuals constantly attempt to defeat the defenses or offenses of another, could easily fill our environment with competing nanomachines. Or more benign forms of surveillance conducted by invisible machines might threaten our current notions of privacy. Do scientists and engineers owe us some duty to foresee and prevent these possibilities?

Scientific Duties and Dangerous Technologies

Consider this: You are a nanotech researcher, trying to wed computing with materials. Your dream is to create molecule-sized robots that will do our bidding, constructing items atom-by-atom, capable of user-generated, customized alterations, and fully recyclable. As you near your goal, you realize the full potential of your new creation, which could just as easily disassemble a human being as construct a cup. You consider not only whether and how evil uses of this breakthrough technology might be prevented, but also whether the potential harms outweigh all potential benefits.

Usually, when discussing the moral implications of various technologies and sciences, we take for granted certain presuppositions, such as 'we can't put the genie back in the bottle', and that ethics is the realm of technologists, not scientists, since scientists have a duty to explore all questions, but technologists have no duty to release every technology. Is it conceivable that these presuppositions are erroneous? Do scientists have duties, regarding especially dangerous aspects of nature, *not* to pursue certain fields of research? Do they share responsibility with technologists who eventually release dangerous technologies? Does this responsibility involve moral culpability whether or not there are any harms that result?

These questions and assumptions deserve a second look. They were the focus of a number of thinkers, including scientists such as Einstein and Oppenheimer, at the beginning of the nuclear age when scientists who had been involved in the development of nuclear weaponry came to oppose the tools they had helped develop. It's an age-old angst, born in *Faust* and *Frankenstein*, involving the inevitable clash of unbounded, unfettered scientific exploration and deadly consequences that sometimes result. Too often, scientists have plodded or lunged along, investigating new means of engineering more destructive technologies, insulated by the concept that science should never be stifled, and that liability for the tools eventually developed because of their investigations rests solely on technologists, engineers, and politicians. But what justifies the disavowal of moral culpability by those in the best position to reveal new and deadly aspects of nature? Is there any moral duty on the part of individual scientists to simply say 'no' to investigating those things whose only or best uses are to cause harms?

The Scientific Firewall

It has long been a staple of ethical debate regarding scientific research that science is open and free, and only engineers need worry about the applications to which they put scientific discoveries. The canard goes: science is ethically neutral, and there is, in fact, an ethical duty to investigate all natural phenomena, so therefore, no scientist need stifle his or her own research. The next step of this argument is to assert that while science ought to be utterly unfettered and free, technologists, engineers, and politicians are both practically responsible and morally culpable for the uses to which any scientific discovery is put. This argument works best with 'dual-use' scientific subjects, like bioweapons, nuclear fission, and fusion.[2] The tremendous possible peaceful uses of thermonuclear technologies argue well for most scientific investigation regarding the underlying sciences. But is this true for all sciences; do they all have 'dual-use' features that insulate scientists from moral culpability when doing basic research?

Consider the recent, real-world example of smallpox. By 1977, a worldwide concerted effort led to the successful eradication of smallpox in the natural world. Its only host is humans, and in the years since its successful eradication, no naturally occurring infection has been documented. This was one of the most successful and heralded scientific and technical enterprises ever. The smallpox virus was virtually extinct, except for some notable stockpiles. The two nations that spearheaded the eradication, coordinated by the World Health Organization (WHO), were the United States and the Soviet Union. Each maintained frozen stockpiles of smallpox samples following the eradication, ostensibly for the purposes of doing scientific research. Then things got out of hand. While WHO and others debated whether the remaining stockpiles ought to be destroyed, some scientists chimed in against the plan. They argued that stockpiles ought to be kept so that further research on smallpox could be done. Some even argued against the eradication of a virus species on moral grounds. The decision to destroy the stockpiles was delayed, and then the stockpiles began to be exploited. There is evidence, including the statements of former Russian president Boris Yeltsin, that during the Cold War, the Soviet Union defied the Biological Weapons treaty and weaponized smallpox, producing it in bulk, and making it more deadly by 'heating' it up, essentially making it less vulnerable to existing vaccinations and anti-viral drugs by exposing it to evolve more robust strains.[3] In the process, stores of smallpox apparently left at least one of the two designated repositories, and now the genie is likely once again out of the bottle.

Once again, in 1999, the world's two custodians of the only known stockpiles of smallpox were on the brink of deciding to destroy the stockpiles (inasmuch as they were believed still to solely exist in the hands of Russian and US scientists) when again some scientists chimed in with a chorus of objections. There were also scientists, some of whom had worked on the original eradication, who argued for the final destruction of smallpox everywhere. In the United States, President Bill Clinton was convinced by those who favored preserving the stockpiles, and the window has now finally closed. The Centers for Disease Control and others working with the US Department of Defense engaged in some controversial studies with smallpox in 2000 and 2001, and successfully created an animal model of the disease, a scientific feat that had hitherto been deemed impossible. This research has since been criticized as being over-hyped, as the animal models that resulted required extravagant forms of exposure before contracting smallpox, making them less-than-ideal subjects for experimentation. The research has further been criticized as being unlikely to lead to any useful discoveries, given that smallpox has been eradicated from the environment and only poses a threat from the current custodians of the virus: the United States and Russia, who could have eliminated it once and for all, but didn't.[4]

In a similar vein, and related to the decision to revitalize US smallpox research, some Australian scientists made quite a stir when they decided to see what would happen if they did some genetic engineering on the mousepox virus. By tinkering with the virus, inserting a mouse gene into the virus itself, they discovered they could defeat any immunity acquired by mice who had been vaccinated, and created a lethal, vaccine-proof mousepox virus with some simple genetic engineering. When US military researchers got wind of this experiment, reported both at a poster-session at a conference and in a journal article, the repercussions for potential mischief with the smallpox virus were obvious.[5] There are obvious ethical questions that arise with both the Australian mousepox experiment and the US/Russian decision not to destroy the last vestiges of the smallpox virus when the opportunity existed. In each case, to differing extents, one might ask whether the risks of the research justified the potential benefits. Weighing the scientific justification against the potential risks of the research seems inadequate, however, to convey the ethical quandary posed by this and similar research. It is a quandary posed by research and development of other technologies, notably in the twentieth century, and that was partly responsible for the development of modern principles of research ethics. The question one might reasonably ask is: Do scientists owe a duty to humanity beyond the relentless, unfettered search for natural laws? The verdict, at least in the realm of bioethics, has been established to be affirmative … there are general ethical duties that outweigh research itself, and that temper behaviors at least when they directly affect human subjects.[6]

The Bioethics Example

It took some terribly visible ethical lapses by Nazi physicians to begin the discussion of codes of ethics governing research on human subjects. The Nuremberg Code was instituted because of the Nuremberg trials, and revelations about the use of concentration camp prisoners for experiments, devoid of pain-relief measures, any semblance of consent (much less informed consent), or any shred of human dignity. The Nuremberg Code served as the founding basis for the evolution of bioethical principles throughout the rest of the twentieth century. Principles such as the right of subjects to provide informed consent before being experimented on, and of being treated with dignity rather than used as mere means to ends, derive from well-known and generally accepted philosophical traditions, but were ignored historically by researchers even outside of Nazi Germany. Well-known examples such as the Tuskeegee syphilis study, the Milgram study (both in the United States), and others prompted the development and institutionalization of bioethical principles in both professional codes and federal and state laws.[7]

Simply put, before the 'common rule', the Belmont Report, and similar codes in European nations specifically applied through laws and regulation the principles enumerated in the Nuremberg Code, human subjects continued to suffer in the name of science. We might speculate as to why scientists would fail to apply commonly held ethical principles, such as truth-telling, seeking consent, and preventing foreseeable harms, but motivations are ultimately not the issue. The fact is, it took creating institutions intended to oversee research on human subjects in order to finally begin to systematically prevent such abuses. It is very likely that many of the scientists over the ages who have misused human subjects in the course of their experiments never intended to cause undue harms, or justified any harms by the potential for greater rewards from their study. A prime example is the completely un-consented to use of a child by Edward Jenner, the inventor of the smallpox vaccine. Jenner's work involved deliberately trying to infect a child, without adequate controls, animal studies, or consent. Fortunately, his hypothesis turned out to be accurate, and the use of cowpox to vaccinate the child prevented his death. Jenner's work saved countless millions, but his ethics was clearly wanting. Such a study today would not have been possible given that animal trials would be useless without a proper animal model for smallpox.[8]

It is likely that Jenner and other scientists similarly situated never meant specific harm, or at least that they justified the potential for harm to a particular subject by the potential for life-saving new treatments for many. What cases like this illustrate, however, is the fact that science has at times proceeded as though ethical concerns were an afterthought, or even completely tangential to the scientific enterprise. Even after the Nuremberg trials, scientists fell into the trap of weighing more heavily the value of potential benefits to be gained from research over individual duties of upholding dignity, providing informed consent, justice, and beneficence. Now, Institutional Review Boards and Ethics Committees provide oversight where human subjects are used in research, and help to give guidance to scientists who might make similar errors. But not all research involves direct use of human subjects. Rather, some research has only potential, future impact on populations, ecosystems, or even humanity as whole. No regulatory body requires vetting of that sort of research.

The example of the development of bioethical principles and institutions intended to apply them suggests that sometimes scientists do not self-regulate when it comes to ethical behaviors. It suggests nothing about motivations, however. It seems likely that ethical lapses are generally caused by lack of introspection, rather than maleficence. This lack of introspection may be exacerbated by the prevailing attitude among philosophers of science and scientists themselves, namely, scientific investigations ought to proceed without limit, and only politicians, technologists, and engineers are to blame for the unethical *applications* of scientific discoveries through technologies. But as is clear in the example of bioethics, and numerous documented examples of ethical lapses by researchers conducting experiments on human subjects, sometimes bad things are done in the name of science itself, well before the application phase of a new technology.

There is a vast and growing literature addressing the ethics of so-called 'dual-use' scientific research, often in the context of the smallpox example, and other bio-security or bio-weapons agents and research. The frame of this discussion has largely included the perpetuation of the notion that 'legitimate' research often has illegitimate uses or consequences. Some have argued from this context that scientists must take upon themselves certain ethical duties and responsibilities, while others have maintained the standard argument that moral culpability lies with those *applying* the research, not those doing it.[9] Looking at the development of bioethics as a field, and considering its institutions and principles, one might ask whether the Belmont Report needs some updating. With a little tweak, we might well fashion a set of principles just as elegant and concise as those enumerated in the Belmont Report, aimed not just at scientists doing research on human subjects, but rather at those whose research impacts *humanity* as a whole. Let's consider this possibility, see how a modified set of Belmont Principles might be applied more generally to all scientific research, and then take up the question of how institutions might then be created that could implement these principles. The discussion is framed with examples, both real-world and hypothetical, and considers also the extent to which some scientific research might never be considered 'dual-use'.

Respect, Beneficence, and Justice

These three essential principles of biomedical ethics frame all reviews of proposed biomedical research involving human subjects. Although based on long-debated principles of ethics in general, and owing much to standard notions of both utilitarian and deontological ethics theories, the Belmont Principles are thoroughly pragmatic, and derived clearly from the most prominent ethical lapses that stoked the report's authorship. They include:

1. respect for persons,
2. beneficence, and
3. justice.

The principle of respect for persons is akin to the Kantian notion that people may not use each other as means to ends, but must treat each other with equal dignity – as ends in themselves. In the Tuskeegee study and other notable lapses of scientific ethics, human subjects were used as means to ends, just as other scientific tools might be, without regard for equal dignity of the subject. The principle of beneficence simply requires that the research on human subjects be conducted with good intentions. It ought to be pursued not merely for the sake of scientific curiosity, but rather to cure some ill, to correct some harm, or otherwise benefit humanity. Finally, the principle of justice requires that populations or individuals who are vulnerable (like children or historically mistreated minorities) must be specifically protected in the course of research.

Debate about the merits and application of these principles continues, but they have also become institutionalized in the form of guiding principles used by governmentally created review bodies that now oversee all research on human subjects in most of the developed world. Despite the philosophical status of debates regarding the Belmont Principles, they are in effect already enacted, accepted as part of the background of all research on human subjects, and devised as a hurdle that must be overcome when proposing new experiments involving human subjects. While ethical lapses still occur, as we have seen with such widely disseminated stories as that of Jesse Gelsinger,[10] we can now gauge the conduct (or misconduct) of researchers involved in these lapses, and educate researchers about how to avoid them in the future. In other contexts, both laws and moral codes do not prevent every harm, but provide contexts for judgments when harms occur. Laws and moral codes still serve to educate, and when agreed upon generally, frame our moral debates over particular acts, intentions, and consequences.

These principles are not unique to the realm of bioethics. They frame many of our everyday acts and intentions, and serve as the basis for both moral and social education and regulation in our everyday lives. Despite their expression and adoption in the realm of 'bioethics', what prevents their application to other fields of investigation? Perhaps it is because the sciences outside biomedicine have had fewer public and noteworthy instances where research has caused visible harms. The deaths or injuries of human subjects used in research typically cause public outrage and provoke action. Research which has no such immediate consequence is unlikely to get that sort of notice. But does this mean that the Belmont Principles are not more generally applicable? If we believe that these principles have no application outside of biomedical research involving human subjects, then we must justify some moral horizon for intentions and acts of scientists. In other words, we would have to justify ignoring the potential for misuse or harms to those not immediately within the control of the researcher, even where those harms might well outweigh or outnumber harms posed to potential human subjects.

To put this into context, let's consider a fictional technology at the center of Kurt Vonnegut's well-known breakthrough novel, *Cat's Cradle*. In this novel, a fictional scientist named Felix Hoenikker discovers a permutation of water that is solid at room temperature. He hides his discovery before he dies, but the secret remains in the possession of his children. Ice-nine possesses the ability to turn any body of water solid given that a single molecule of it will 'teach' all other molecules next to it to become ice-nine, creating a chain reaction that freezes any body of water with which it makes contact. It does the same to any water in an organism's body if ingested. The fictional ice-nine is clearly a terrifying scientific discovery. Vonnegut based the character Hoenikker on the Nobel Prize-winning Irving Langmuir, whom Vonnegut knew through his brother's association with Langmuir at General Electric. Of Langmuir, Vonnegut said: '[he] was absolutely indifferent to the uses that might be made of the truths he dug out of the rock and handed out to whomever was around. But any truth he found was beautiful in its own right, and he didn't give a damn who got it next'.[11]

In the book, ice-nine inevitably gets released into the environment, essentially bringing about the end of the world, all life on it, and all but a handful of people who manage to survive. The research on and discovery of ice-nine would never have fallen under the purview of bioethical principles as enunciated in the Belmont Report. While ice-nine is fictional, smallpox is not and it poses many of the same questions, real-world threats, and ethical challenges as that posed by Vonnegut's book. Is the beauty and truth of science justification enough to investigate even the deadliest potential technologies or discoveries? Are there ethical principles that bind individual scientists in the absence of regulatory institutions? Can the Belmont Principles be extended to scientists doing research only *indirectly* involving human subjects, where the potential effects of an avenue of study impact *humanity as a whole*?

Extending the Moral Horizon

Most arguments concerning the morality of certain types of basic research focus on issues of dual-use and unintended consequences. These arguments concentrate on the distinction between 'legitimate' scientific investigation versus unethical uses of the research. As discussed above, this presupposes that scientific research is always in a different moral position than the application of that research. What justifies this assumption? Do scientists enjoy a unique position occupied by no other fields or professions? Let's examine some reasons why they might before considering whether scientific inquiry itself, prior to its application through a particular technology, may ever confer moral culpability.

Some might contend that scientific inquiry alone can never confer moral culpability because inquiry is personal, a matter of conscience, and subject to no restrictions at all. Limiting inquiry in one realm might lead us on a slippery slope of censorship, thought-police, and other clearly unsavory interference with free thought. We don't want regulators to prevent scientists from doing legitimate research, looking over shoulders to police scientific investigations, and preventing the acquisition of knowledge about nature. Indeed, governmental interference with scientists' research has provoked the wrath of both scientists and the public, especially when done in the name of particular ideologies.[12] Let's take it as a given that this sort of regulation is tricky at best, Orwellian at worst. But just because we don't want government or regulators overseeing the actions of an individual, doesn't mean that that person's actions, or even intentions, are free from moral scrutiny. We often and comfortably make moral judgments about conduct and intentions that have no *direct* effect on others, even when we don't tolerate or desire any institutional regulation. Yet there are still strong arguments supporting the notion of unfettered scientific inquiry, devoid both of governmental and self-regulation.

Science doesn't kill people; people with *technologies* kill people. Of course, this is a perversion of the US National Rifle Association motto: 'guns don't kill people; people kill people'. There is some sense to this. Artifacts like crossbows, rifles, and nuclear weapons cannot be morally culpable, only people can be. In the name of greater personal freedom, of both conscience and property, governments ought not to restrict inquiry into, or ownership of, dangerous items. The law, codes of ethics, and public and private morality are well-equipped to deal with unethical uses of artifacts, so the principle of maximal freedom requires that we allow not only inquiry into, but also possession of knives, rifles, and nuclear weapons (at least for certain nations). But this is not quite the case in practice, and we do tolerate restrictions on owning certain artifacts. Thus, in the United States, even the most ardent gun aficionado cannot legally own a fully automatic weapon, to say nothing of a tactical nuclear bomb. Moreover, we do not restrict thinking about, and inquiring to an extent, laws of nature generally. Indeed, many of us would consider it immoral to restrict such thought or inquiry as an intrusion into matters of personal conscience. But does this necessarily imply that while the freedom of personal conscience enables us to think about and inquire into all the universe's natural laws, taking any and all further steps must escape moral judgment?

Take, for instance, the problem of child pornography. Do we hold a pedophile morally guiltless just because, while he might have amassed a collection of pedophilic literature or cartoons, he or she never actually abused a child or contributed to the abuse of a child? Notions of free speech and conscience might protect that behavior, but we are willing to judge certain further positive acts relating to pedophilia as not only morally blameworthy, but worthy also of legal culpability. Intentions matter, even when intention alone is not enough to spur public regulation. Stated intentions matter more, and even when they do not rise to the level of legal culpability, they may spark appropriate moral indignation or outrage. And finally, positive acts based on intention matter even more, and can provoke appropriate moral indignation, outrage, or public recrimination. The pedophile who possesses photos, even while he or she might not have directly contributed to a harm, has indirectly done so and our moral outrage and legal repercussions grow accordingly.

The case of the pedophile might make us reconsider the notion that, while all thought and conscience should be totally free of external restriction, both are nonetheless immune to *moral judgment.* Similar cases may be made about individuals in both their personal and professional capacities who hold intentions, and even take positive actions without direct consequences or harms, yet that invoke some moral culpability. Do we hold the businessperson who thinks about the social or personal consequences of his or her actions in the same regard as one who does not, even where no real difference accrues to customers, colleagues, or society?

I argue that when considering the ethics of scientists, we must not only look at regulations, laws, and codes used to review or punish their actions, but also consider intentions and motivations with an eye toward education. Moral training of scientists, as with other professionals, presupposes not only that we wish to keep them from breaking laws or running afoul of professional codes of conduct, but also that we wish to help develop moral insight that can guide behaviors.[13] Take an example from another profession whose members affect peoples' lives daily, with sometime dire consequences. Even where an attorney, for instance, injures no one by his lies, the fact of the lie alone ought to concern us. Both in their individual and professional capacities, people who lie are not to be trusted and deserve our moral judgment. In professions like engineering, science, medicine, and law, moreover, the consequences of actions taken with ill-intentions matter much more to clients, colleagues, patients, and society as a whole simply because the potential for harms is so great.

We should take account of a broader moral horizon when considering the ethics and morality of scientists, just as we do with other professionals, and ask whether and when intentions and positive actions on those intentions trigger an individual duty to refrain, and subsequent moral judgment by others, even where the law or regulations ought not to be invoked. All of which brings us back to the scenarios presented at the outset, and the problem of ice-nine as described by Kurt Vonnegut. Do the principles of beneficence, dignity, or justice provoke any ethical duties for scientists while inquiring into natural laws? Are these duties, if any, different in so-called 'dual-use' cases than for instance in the case of nightmarish scenarios like ice-nine?

Smallpox, Ice-nine, and Nanowares

Almost anything can be considered 'dual-use' if we want to be technical. A nuclear bomb could be used to level a city, or to create a canal. Smallpox research could be used to develop new cures, new therapies, antiviral medications, or new biological weapons. Even ice-nine could serve a dual use, providing ice to skate upon in the middle of summer, or destroying the entire ecosystem. For that matter, the most seemingly benign inventions could, given sufficient creativity, be put to questionable or immoral uses. Printing presses can produce great works of art, or hateful screeds. The dual-use debate, then, may be a bit of a red herring. We are concerned with the ethical implications of certain types of scientific research, and the capacity for a certain discovery or technology's dual-use is not what matters. Instead, we should ask under what conditions a scientist ought to refrain from investigating some aspect of nature, and under what conditions he or she ought to disseminate certain knowledge, regardless of whether the science in question has both a beneficial and harmful use. Let's reconsider the issue of smallpox and its near-eradication.

The global public-health initiative to eradicate smallpox was nearly successful. Its final success was abandoned, and now, despite the fact that smallpox does not exist 'in nature' it still exists, and poses a real threat to humanity. That need not have been the case. Because smallpox has no other vectors for its survival apart from human hosts, when it was finally eradicated from all human hosts its extinction could have been guaranteed. But for the fact that the United States and Soviet Union insisted on maintaining stockpiles of the virus, we would not need now to worry about the use by rogue states or terrorists of stolen quantities of smallpox. But for the efforts of the United States and the Soviet Union to 'study' the use of smallpox in bio-warfare, and the mass production and subsequent loss of control over the remaining stockpiles of smallpox virus under Soviet science, smallpox would be but a distant memory of nature's capacity for horror and destruction. Scientists cannot be held immune from the moral implications of having preserved stockpiles of the virus. Some made arguments based upon the value to science posed by continuing study of the nearly extinct virus. Their arguments won the day, even if there may have been ulterior motives on the part of the two sponsoring governments maintaining the last known samples. Do any ethical principles mitigate against either the active encouragement of or complicity in the decision to retain the last remaining smallpox samples?

Let's consider first the Belmont Principles. As it turns out, one of the big obstacles to conducting any legitimate science with smallpox is that it has no animal vectors. The Nuremberg Code, and subsequent codes of biomedical ethics, requires that research on human subject be preceded by animal research. To do useful, beneficial research using smallpox would require a human subject, and no researcher could ethically employ human subjects in smallpox research. Not only would the danger be too great, but without first doing animal research, no human subject research could be approved. In the last-ditch effort to save the smallpox stockpiles in the United States in 1999, researchers proposed that a useful application of the smallpox samples was in attempting to find an animal model for the disease. Toward this end, researchers exposed monkeys to massive doses of smallpox until they finally sickened the subjects. Nearly every monkey exposed died without developing human-like symptoms of the disease. But a couple monkeys developed responses similar to human smallpox. This was written of as a triumph in smallpox research, and for some has justified the maintenance of the smallpox stockpile. Finally, a potentially useful animal model of smallpox infection may exist which now justifies maintaining the stockpiles so that further research can be done. And all of this is further justified by the very real potential that smallpox, while no longer a natural threat, is a threatened potential agent of bio-terrorism.[14] In this context, what are the implications of the decision to preserve smallpox, considering the principles of *respect for persons, beneficence,* and *justice*? Does an extended moral horizon alter our view?

If we consider that the subjects of the smallpox investigations (conducted in part to justify continuing to maintain smallpox stockpiles) include not just the monkeys that were infected and ultimately died, but also humanity as a whole, did this experiment satisfy the Belmont Principles? It would arguably meet these principles *if* smallpox remained a natural threat. The dignity of individual humans was not infringed. No individual was treated as less than autonomous or deserving of equal dignity. Moreover, if smallpox were still a natural threat, then presumably all experiments would be conducted with the goal of treatment or, as was the case before 1979, eradication. And finally, the principle of justice is satisfied as long as no vulnerable populations were treated less than equally in the course of the experiment. But if we consider the implications of the experiments in the context of a disease that could historically have been eradicated completely, then we can be more critical of the intentions of the scientists and their decisions to take part in the research.

Let's imagine, since smallpox had been eradicated from the natural environment and only posed a threat from intentionally maintained stockpiles held by humans, that smallpox and ice-nine pose nearly identical risks, and are similar technologies. Ice-nine, like smallpox, posed no natural risk in Vonnegut's book, but only posed a risk as a human-devised technology. The dual-use argument that might justify experimenting with ice-nine breaks down in light of its artificial nature. Moreover, the potentially catastrophic results of an accident involving ice-nine (namely, the total destruction of the biosphere) argue in favor of a positive duty not to investigate it beyond mere surmise or theory. Neither beneficence nor justice warrant investigating ice-nine. We might argue that beneficence argues in favor of investigating smallpox because we worry about terrorist uses of it and need to devise treatments. All of which is recursively self-satisfying, because we would not have had to worry about this had scientists done the right thing to begin with, and supported its ultimate destruction. In the world of *Cat's Cradle,* we could similarly argue in favor of ethically pursuing ice-nine research *only* in a post-ice-nine-apocalypse environment.

An argument that is often used to justify these sorts of scientific inquiries is that 'someone will devise the technologies, and employ them harmfully – eventually. Thus, we should investigate these things first (because we have good intentions)'. Of course, this reasoning justifies investigating any and all science and technologies, no matter how potentially destructive or threatening to humanity or the environment. But it presupposes (a) that the investigators doing the work have good intentions, (b) that the technology or discovery would eventually be carried out by others, and (c) that once discovered or applied, it can be contained. Let's assume that, in fact, ice-nine, or smallpox for that matter, will come into the hands of individuals or groups with less-than-good intentions. Will discovering it or investigating it now do any good? In the case of ice-nine, clearly the answer is no. In the case of smallpox, beneficence would argue for the research if for some reason we believed that smallpox could not be contained with existing technologies. If, for instance, we believe that the Australian 'mousepox trick' could be applied to smallpox, then devising ways to treat genetically altered mousepox might be an act of beneficence. But without an animal model for similarly altered smallpox, we'd need first to try the 'trick' on smallpox. Again, we have a recursive, self-fulfilling justification to pursue any and all research, including on any devastating, horrific, or deadly technology one can think of. Moreover, there's always reason to question whether one's own motivations will always be pure, or that a technology will always remain in one's control or contained.

The 'Eventual' Fallacy

The 'eventual' fallacy justifies any investigation, and scientific inquiry, no matter the potential consequences. It fails if we broaden the moral horizon offered by the Belmont Principles to include *humanity as a whole* when we are considering sciences and technologies posing no natural threat. Implicit in bioethical principles is some utilitarian calculus. Science proceeds not in a vacuum, but as a socially devised institution. It is conducted by professionals, with funding from mostly public sources, and with relative freedom under the auspices of mostly academic environments. As a largely public institution, and as the beneficiaries of the public trust and wealth, scientists must consider the consequences of their inquiries. They are not lone, mad scientists, hunched over Frankenstein apparatus in private attics. Nor is worrying about the possible existence of Dr. Frankenstein sufficient to warrant all inquiries. Rather, science must be free to inquire into any and all of nature's mysteries, but scientists must also be aware of being beholden to a world at large, bound by concerns about consequences of their research, and ultimately dependent upon public support for their institutions.

The 'eventual' argument makes sense when the risks posed by investigating a deadly thing are outweighed by the likelihood of that deadly thing being discovered and used by others combined with the potential of a scientific investigation developing a plausible protection of the public at large. So, roughly

R = risk,

L = likelihood of independent discovery and use, and

P = potential benefit from scientific investigation *now*.

If $L + P > R$, then a scientist can make a moral case for pursuing an investigation into something posing a large, general risk. Otherwise, there is simply no moral justification for further inquiry. Taking ice-nine as an example

R = 100 (near-likelihood of worldwide catastrophe if released into the environment),

L = 99 (being generous, on a long enough timeline, this number grows to 100), and

P = 0 (there's no 'cure' for ice-nine).

Or taking smallpox research (now, as opposed to before the eradication)

R = 90 (smallpox could escape and cause enormous human devastation),

L = 70 (there's a chance that Russian stockpiles have already made their way into others' hands), and

P = 10 (we already have smallpox vaccines that work well, but maybe we can develop vaccines for other strains or genetically modified versions).

Note that the value of P changes as the likelihood of potential independent discovery changes because of the temporal *caveat*. Thus, it is right to inquire into the state of scientific knowledge elsewhere. However, this imposes an additional burden not to increase the value of L by *disseminating* knowledge that leads others to dangerous knowledge where there is no supervening imperative due to potential benefits from the knowledge.

The 'eventual' argument changes over time, and depending upon actual conditions in the world. If, for instance, we know that some rogue state or terrorist group has been experimenting with smallpox, then the calculus changes. It changes even more if we can identify the nature of those experiments and thus target scientific inquiry to a specific threat. But a generalized threat posed by the potential of someone acquiring knowledge or technology somewhere at some time means that this calculus requires scientific caution. For sufficiently deadly inquiries or applications, scientists should perceive a duty to refrain at least from disseminating certain types of knowledge, if not necessarily from theoretical inquiry alone (unless that inquiry may reasonably be justified by the above calculus). The 'eventual' fallacy is committed by simple recourse to the truism that over an infinite time-span, all natural truths will be discovered, and all potential uses and misuses of technology will be developed, so present research on any science or application of technology is morally justified.

Implications for Institutions

Unlike the Belmont Principles, which could be used to guide the development of regulatory institutions, the expanded ethical horizon I have argued for above requires individual responsibility on the part of scientists. The calculus proposed must be employed by scientists before they ever get to the point of disseminating their ideas. It is a personal, moral responsibility that must be cultivated. Nonetheless, encouraging the development and adoption of these principles, and adopting the notion of a broad horizon of scientific responsibility (encompassing not just individual human subjects, but also responsibility toward *humanity in general*), can best be encouraged through new institutions. Legal and regulatory bodies ought to devise these institutions both within and among sovereigns. Professional organizations as well ought to embrace and adopt ethical training of their members, understanding that scientists are citizens of broader groups whose funding and support they require. Education in principles, not just of scientific integrity, but also social responsibility, ought to be developed and embraced. Currently, scientific integrity and ethics are taught only in the briefest and most superficial manners, and are not generally necessary for any scientist not doing research on human subjects. But in light of the potential for sciences and their technologies to be used for harm, and given the scale of some of these potential harms, more general education in science and morality should be required. This is especially true where the potential impact of a particular science is great, as with nanotechnology, genetic engineering, and similar technologies.[15]

As discussed above, scientists are generally beholden to public funding, at least to some extent, and just as many governments now require some minimum training in the core bioethical principles of Belmont and its offspring, so too could grant funding in technologies like nanotech and genomics depend on some minimum education in ethics.

Besides education, the principles and proposed calculus of risks, harms, and benefits could be used in *post hoc* analyses to determine culpability where scientists release dangerous sciences or technologies which actually cause harms. Just as medical doctors were culpable in the Nuremberg trials, so too might future scientists be morally and legally culpable for the apocalyptic (or even slightly less so) repercussions of their negligence or recklessness. Of course, *mens rea* must be considered, but merely citing the 'eventual' fallacy will not suffice to defend all scientific inquiry and its resultant dissemination, either through publication or technology. Scientists must not only have a sense that they are morally culpable for the uses of their discoveries where they understand the risk–harm–likelihood calculus, but they must also be liable to be held culpable where harms result from their acts, and where they possess a culpable *mens rea.*

Just as governments take it upon themselves to fund and advance research and development, both out of scientific curiosity and as a way to grow economically, so should they adopt the responsibility to educate scientists to be better citizens. As taxpayers provide for investigations into nature's truths, sometimes with no potential for economic benefit, they must also be considered as beneficiaries or targets of the fruits of scientific inquiry. An expansion of the Belmont Principles might include recognizing *we are all human subjects of certain inquiries.* Where discoveries possess the potential of great harms, environmental catastrophes, mass extinctions, or worse, the collapse of an entire biosphere (as with ice-nine, or 'grey goo'), scientists must take it upon themselves to measure their aesthetic appreciation of truths in themselves with gravity of worst-case repercussions. Institutions and regulatory bodies must encourage this, and provide guidance in the form of practical moral education of all scientists, not just in medicine and bioethics, but for all fields of inquiry.

Teaching ethical principles to scientists need not stifle research. Nor does it imply that scientists must watch their thoughts. They need not restrict their thoughts, but they ought to guide them. Minds should be free to explore all possibilities, but the context for inquiry must always be considered to encompass something broader than just the institutions of science. Where one realizes grave or catastrophic implications for a particular path of inquiry, one *does* owe a duty to those on whose behalf they are musing, and who would inevitably become the target of resulting catastrophic technologies. Just as any of us may privately muse about acts of horror or violence, we assume greater duties as we begin to discuss, plan, or execute those acts. The same must be true for scientists, as in any other public profession or private life. Respect, beneficence, and justice apply not only to human subjects of particular experiments, but also more generally to humanity as a whole. The result of all this should be better trust of scientists and their profession, and a greater realization on the part of scientists that their work proceeds through mutual trust and appreciation between scientists and the public. In the end, we all should benefit as scientists begin to realize their duties are personally held, and broadly applicable. When faced with the choice to inquire into something whose only or most likely application is harmful or deadly, scientists should have the moral strength, educational background, and public support to refuse in light of ethical principles generally accepted, well considered, and backed by strong public institutions.

Does the Future Need Us?

That all depends. The world is indifferent to our comings and goings, and species have perished before. Carl Sagan mused that perhaps the reason we had not yet been contacted by alien species was due to some law of development by which, when a species attains the capability to reach out physically beyond the stars, it also develops the technical means (via artifacts like nuclear weapons) to destroy itself. Having attained those technical means, he surmised, most species cannot help but drive themselves into another dark age, or become extinct. We would like very much to believe this won't happen to us, and that the better angels of our natures will be summoned in time to avoid our self-extinction. Whether we escape our extinction at the hands of runaway technology will very much depend upon the individual intentions of a great many actors. It cannot depend upon the actions of states, as states have not only been notoriously bad at regulation and containing dangerous technologies (which they often appropriate for their own wartime uses), but *cannot* technically contain the twin technologies of synthetic biology and nanotech, just because of the nature of these technologies.

The question is: Do we need the future? If we do, and if we wish to see the utopian rather than dystopian nanotech future unfold, and all its various promises met, then we need to make conscious efforts to act as though our present decisions matter, and the wrong ones may be disastrous. Curtailing certain research, or self-regulating and community policing, all have roles to play in ensuring that our technology does not harm ourselves and others, or irrevocably alter our environment for the worse. Recognition of general and specific duties to others, and across time, requires behaviors to be moderated despite a generally recognized and accepted duty to science itself. These duties impact both scientists and innovators, as the subject matter of the two overlap, and those with the closest contact with nascent technologies and sciences are best positioned to anticipate and avoid harms.

It is easy and tempting to brush off apocalyptic visions as the musings of apparent neo-Luddites and science fiction writers, but sci-fi dystopias sometimes play out as envisioned. *Fahrenheit 451* can be read as a warning of a future that is nearly here, and in both our popular culture and the political sphere, we are aided by even fictional scenarios in planning for real futures, or avoiding them. Crichton and Joy's nightmarish futures *should* be avoided, and we have clear duties to avoid them where possible. Utopian visions of nanotech similarly serve as beacons for where and how we can direct our intentions, and steer ourselves toward a particular destination.

We may also seek to avoid, as we build or modify institutions by which the future of the technology will be constrained or liberated, certain *social* dystopias. Stephenson's *Diamond Age*, as much as it portrays a technological utopia, also warns of the social dangers of monopoly, suggesting that central control of the technology will lead to new class divisions and both unnecessary and undesirable scarcities. Extending the current IP paradigm over nanotech is not only pragmatically unworkable, ethically unjustified, and economically inefficient, but also anathema to the (small 'l') liberal goals we promote in our politics. In *The Diamond Age*, it is overcome through technological and political revolution of a sort. We can start on the ground floor, ensuring that the technological means to ensure democratic technologies are built into our institutions as well as the technology. We can adopt the perspective that openness now will help us prevent tyranny later. History proves this true.

CHAPTER NINE

The Final Convergence

Intellectual property (IP) was a noble experiment, and it may have been the cause of some of the growth and development we have enjoyed in the last century or two. But its days are now numbered. The end of IP is written into the structure of an ongoing technical and scientific revolution. Nanowares will be the final merger of numerous seemingly disparate technologies into a single, overarching technological infrastructure. This revolution, when fully realized, will alter our fundamental relations among each other, to the products of our creativity, and to the 'natural' world. No longer will many of the forces that have driven prices in markets function as they have. Scarcity, in the best version of the nanoware future, will cease. This future is being prepared for by many who are trying to develop it.

I have been trying to make a case, ever since I first examined the nature of cyberspatial goods, that IP regimes do not reflect reality well, and that they can and should be altered (or abolished) to serve our needs, and to better embody both logic and pragmatic ends. The *technology* is revealing that our preconceptions about the nature of our technical artifacts, as applied through existing IP regimes, are fundamentally flawed. The case is being made for us by both the inadequacy of current legal norms to deal with emerging technologies and the recognition of innovators, who now consciously skirt those regimes, and are instituting and embracing new means to promote and to protect innovation. If we are to pave the way for the full, beneficial, and equitable realization of nanowares, then we must finally declare IP as we know it to be dead, and move on. Let's review what our examination has revealed, and explain why the death of IP means opening a new door for the future of technological and scientific progress and economic prosperity if properly navigated.

Nanowares: What Are They, Really?

I have attempted to use a single word to connect several different paths of current technological approaches to the same goal. The sciences have been pushing technology inevitably toward a convergence, as the fundamental goals of science – better understanding, prediction, and control of nature – have been leading to the development of the technologies that underlie nanowares. The emergence of essentially programmable physical goods will complete the development of two major tracks of technological development. On the one hand, computerization and information and communication technologies (ICT) have involved ever-increasing abilities to manipulate information, and on the other hand, better, more precise, and miniaturized manipulation of the physical world (necessitated in many ways by the underlying technologies of ICT) has involved ever-increasing control over our manufactures or artifacts. The convergence of information technologies and manufacturing means fundamentally rethinking our connection to our artifacts, as discussed in the chapters above.

One exciting feature of these converging technologies is *how* they are finally converging. I have specifically included in the analysis of nanowares technologies that are clearly not nanotechnology. There are two reasons for including in discussions about nanotech and innovation the grassroots development of localized or microfabrication technologies. The first is that those who are behind the emergence of these technologies have their theoretical roots in nanotech. The second is that the tools and infrastructures that they are developing capture many of the same issues as those that will be presented by the full realization of molecular nanotechnology (MNT). The goals of enabling the complete accessibility of all potential means of production, and of liberating fully human creativity from the traditional demands of capital, are implicit in the Fab Lab, RepRap (and other) micro-manufacturing efforts, and will be the ultimate inevitable result of true MNT. Because I (and others) see both streams of development as meeting at some future point, and motivated by the same ultimate goals, a single term 'nanowares' has been used throughout this argument.

We should treat all of these emerging and converging technologies as a kind, and look for ways we can precipitate their smooth arrival and realize their full potential. Ultimately, I do not think there is any real danger that governments or others who foolishly seek to keep alive current IP paradigms, and extend them to nanowares, can succeed. In fact, it is encouraging that while some of us try to justify the theoretical underpinnings of a movement, the movement itself marches inexorably onward, propelled by the general acceptance of its tenets. IP has been breaking down for some time as a result of half of the two-track movement toward true MNT. Even as the patent profession (really, the patent industry, in which patent lawyers, courts, and other related professionals are assured of profits by the mere filing and litigation of patents) seeks to bolster their floundering field against the tide of radical change brought about by ICT, those who are innovating in software have learned how to work around the legal impediments to innovation, and work out among themselves a more just spectrum of rights and expectations. This is happening as well in nanowares. It is part of the nature of the technology that openness and free markets will drive innovation in the field, and it moves those working in it at least away from excessive upstream patenting, and at most, toward open source paradigms.

In both the grassroots nanowares approaches to microfabrication (including desktop fabrication) and in the realm of synthetic biology, where bio-bricks approaches ensure free and open access to the foundational research and components, researchers and innovators alike are taking matters into their own hands and preventing patents from impeding these emerging sciences. The same trend motivated both industry and individuals to disclose rather than enclose genes in the race to decode the human genome and thereafter. Nanowares are thus part of a spectrum, and the continuation of a trend. The patent industry will grow louder, and look to governments to secure their domains, even as those who are doing much of the work in both the science and innovation work steadily to keep this expanding domain open.

In many ways, this book is not a call for action, but rather an attempt to theoretically explain trends that will occur anyway. It is descriptive rather than normative. Nanowares will be open, eventually, and IP will continue to break down, with or without philosophers making the case that it should. But some remain unconvinced, and innovators and scientists are still at the mercy of states, as well as market forces, that sometimes work at cross-purposes, and that are often impelled by habit rather than reason to continue to support dying methods and ideologies. Even while I am certain that the seeds of revolutionary change in our thinking about the natures of artifacts are embedded in the technology itself, we should be mindful of the pernicious ability of those whose domains are threatened, even by the inevitable, to attempt to secure their monopolies.

I have argued throughout my informal IP trilogy that where institutions or norms conflict with grounded principles of justice, we must seek justice. This is an ethical argument, and suggests that the forces I describe, and which are moving technology and science, are *good*, and that attempts to impede that development are the opposite. You may well accept the trends I have described, their economic efficiency, their inevitability, and their benefits for innovation without accepting the ontological and ethical arguments that pervade this text, but I have tried to point out the shortcomings of mere utility as a foundation for action. Let's look briefly at why ethics and justice ought to finally motivate us to embrace nanoware without IP, and why ethical foundations ought to be part of our thinking in science and technology in general. There are sound historical reasons to believe that some things are better than others, not just instrumentally, but also *fundamentally*.

Ethics and Innovation

If science and technology never had any impact on rights or duties, and if IP law did not impede some of those rights and duties, then philosophers could avoid confronting the ethical issues raised by new technologies like nanowares. But each new technology presents us with a range of new considerations regarding the rights of scientists, innovators, and the public, and the proper role of states in ensuring justice. It is a modern trend to incorporate earlier in the development of new technologies ethical considerations that are anticipated to pose impediments or lead to harms. As discussed above, most of the discussion regarding nanowares has focused upon potential harms posed in the realm of safety, security, public risks, and environmental impacts. All of this inquiry is warranted, given the experiences of harmful technologies in the past, and we have delved into them briefly above. But the argument I have been making in this text focuses primarily on IP, and how the current IP regime threatens the development of the technology itself. While I am mindful of, and concerned about, potential physical harms arising from nanowares, and indeed support inquiry into the general ethical duties of researchers and innovators in avoiding those sorts of harms, I am most concerned with the implications for justice arising from the application of current IP norms to this and other technologies.

We are not, as individuals, entitled to the fruits of innovation. I do not think there is a duty, either, for anyone to innovate. Nor do I think that justice requires the free dissemination of physical goods that result from innovation to either the needy or the wealthy. But I do think that states ought not to impede either individual human creativity or markets for the fruits of that creativity. There may well be individual, moral duties to contribute to society, and to not harm others consistent with Mill's oft-spoken about (and incorporated herein) 'harm principle.' I would certainly encourage innovators to innovate, to share the fruits of their innovation openly, and to enable others to improve upon their own innovations. Doing so would certainly be what ethicists consider to be supererogatory goods, or praiseworthy (morally speaking). But this is not an argument for achieving some moral utopia through adopting socialist principles such as the famous maxim: 'to each according to their needs, from each according to their ability.' Instead, it is a call to let markets sort out prices, free from state interference, and to allow human creativity to thrive without creating artificial barriers in the form of state-sanctioned monopolies.

I do not mean to imply that an utterly free market in innovation will result universally in 'the good,' ethically speaking. Rather, it is clear from history that people will do bad things, intentionally or unintentionally, and that states have sometimes tried to step in to rectify harms in some cases, and succeeded sometimes. The law and the state have played roles in marshaling public resources to great good in the past, and have in fact helped us to achieve our current level of innovation and wealth through various means, including through the allocation of resources to both science and industry. It is not *per se* unethical that we should make such democratic choices about allocation, control, regulation, or preference where those choices are truly democratic, and where harms might be avoided with certainty. Neither is it *per se* necessary that states do any of these. Looking at past examples, as we have a bit throughout the above argument, there are clear instances of state failures, in both allowing for and in some cases precipitating harms. Moreover, there is an even clearer history of state error in its actions regarding markets. There is as yet no good evidence that attempts to influence markets, especially in the development and dissemination of new technologies, are necessarily successful. Such attempts are also arguably bad, in a moral sense, when they involve impediments of natural rights (to life, liberty, or property, for instance). I believe that nanotechnology presents us with an important choice, and that many who are involved in this nascent field have already made that choice. The choice is: Do we wish to adopt failed IP policies in nanowares, or pursue new paths to liberate both the science and its development through technology?

We have other choices to make as well. For instance, should we pursue informal policies, or create institutions that help us to track dangerous materials, precursors, or final products that enter the stream of commerce? Are there some sorts of technologies that need to incorporate limitations, for instance on the potential negative consequences of unhindered self-replication? Should scientists and innovators choose to abide by certain ethical principles in general? All of these choices have ethical implications. The common thread of everything I have tried to demonstrate, and the overarching thesis of this book, is that it is the nature of each of the above as *choices* that is most important.

Nina Paley, *Mimi & Eunice* (cc) creative commons

Choosing to Do Better

With the move away from classical scholasticism, and the modern embrace of Enlightenment liberalism, we have abandoned largely the notion that the good is that which must be done because of some commandment handed down by an authority. Rather, the turn of ethical theory, and of institutions, has imbued in individuals a greater role in choosing the good, directing their intentions, and in accepting personal responsibility for their actions. No other system could suffice for a complicated world with ever-expanding realms for decision-making, and without historical guidance for directing our intentions and actions to new, unanticipated situations. There is no clearer argument against a scholastic approach, and in favor of liberalism, than the trajectory of science and technology since the Enlightenment began. By liberalism, we mean an approach to questions that embraces empirical evidence, and that is open to reform. Moreover, liberalism centers power not in the state, or in some higher authority (like a sovereign or deity), but rather in the individual, from whose consent the just power of a sovereign must constantly derive. Political liberalism recognizes the contingent nature of the state's authority, and the ultimate authority of citizens over the state. It also recognizes that the methods of the sciences, by which truths are forever amenable to falsification, are applicable too to the mechanisms of state power. But classical liberalism of the variety we have inherited from Locke and others also recognizes certain immutable principles of right that must guide and limit the state, including inviolable rights of individuals to maintain their liberty, property, and lives. These are like axioms in mathematics and the sciences, and while they are open to falsification, the burden of proof necessary to overturn an axiom is greater than other contingent truths.

Out of liberalism, and the methodology of the sciences, came two great historical attempts to devise a universal ethics. These two schools contend today. Kant's deontology and Bentham's utilitarianism stand as two distinctly different approaches to ethical problems. While I clearly favor one, our agreeing that deontology is best is not necessary nor sufficient for us to reach the same conclusion about the role of the state in innovation, or about the ethics of IP in general. Utilitarian arguments might very well be made that suggest that IP laws impede general utility. But that is neither here nor there. What is critical is that both of these major, post-enlightenment theories of ethics present us with the capacity and responsibility as individuals to make choices, reasoning out the ethical consequences or categorical duties directing those choices, rather than blindly accepting the wisdom of past generations, sacred books, or other authorities.

The currently accepted wisdom about IP should not be blindly accepted. It must be challenged. If it is supported by empirical evidence, then so be it. If IP does not accomplish the aims it was intended to, then based solely upon a utilitarian calculus, its harms measured against its benefits must impel new choices. Or, if there are categorical duties, natural rights, or further considerations outside of a utilitarian calculus that suggest that IP schemes lead to injustice, then we have more bases for choice. In any event, the fact that we actually have choices, and that we need not accept as immovable the current status of institutions regarding innovation and IP, is what sets us apart as members of liberal democracies. In the case of IP, we need not even change the law. We can choose to develop wholly new institutions and adopt them, even without the intervention or support of states. This is truly liberating.

The above has been an argument that the choices currently being made by whole categories of researchers and innovators are soundly based not just upon pragmatism, but upon justice. Informal and formal institutions exist that *already* enable those who wish to innovate outside of standard IP norms to do so, and to further thwart attempts by those who would lock up information, potentially impede scientific and technical progress, and steadfastly hold to outdated dogmas. It is not just fortunate that many of those spearheading the science, and building these technologies, are choosing to do so in the open, and without IP. It is just.

As I wrote earlier, this is not so much a work of normative ethics as it is one of descriptive metaphysics and ethics. I am documenting trends that I suggest are naturally impelling those who pursue nanowares to adopt open innovation and to reject standard IP. Moreover, they are doing so based upon sound, and ethically justified choices. Many of those connected with what I have called the 'patent industry' argue that such choices are foolish, that they will slow innovation, and that they deny rights to innovators. We should critically examine and question the motives behind such opinions and advice, and then choose, based upon reason, not fear, how to pursue nanoware science and innovation. I believe I have made a case throughout this book that choosing open innovation, and rejecting IP, is wise, ethical, and pro-innovative. The evidence, and the analysis I have offered, suggests that the end of IP is coming anyway. It has been slowly breaking down, its weak foundations revealed by emerging technologies since the advent of ICT, and completed with nanowares. What we are faced with now is a choice, and therein stands the ethical challenge. Will we choose to retreat to the safety of accepted dogma, or brave the future with new models, new approaches, and a greater respect for justice and right, and their necessary relations to science and technology?

The Logical Necessity of Open Innovation

As we have briefly discussed, and as numerous other scholars have noted in much greater depth, science has proceeded apace in the past, and innovations have thrived, even absent explicit state-sponsored monopolies. The past seventy years are an aberration of sorts. Spurred by the emergence of cold-war paradigms of 'big science,' academic institutions, researchers, corporations, and the public have grown used to a mode of operating that seemed for a time to propel us into a brave new age, and even got us to the moon and back, but which no longer seems either workable or necessary. Many have argued, and I agree, that the trend toward the commodification of 'upstream' science contradicts the scientific 'ethos.' As early as 1942, Robert Merton argued that science is an inherently democratic institution, and embodies, when properly pursued, the following values: universalism, communalism (which he unfortunately called 'communism'), disinterestedness, and organized skepticism. There have been many lengthy debates about the adequacy and accuracy of Merton's appeal to a scientific ethos, but the sentiment captured by this list echoes some good, pragmatic values that are certainly a *part* of most scientific inquiry, and that are arguably necessary for scientific progress.[1]

Science progresses through transparency, and tends to become pathological when it is obscure. Scientific progress can only be tested by universal acceptance of its methods and axioms, and communally organized research groups sharing their findings and laying them bare to falsification. Scientists must be detached from the truth (or profitability) of their findings, because self-interest tends to promote lies or delusions. Scientists must remain always skeptical of their findings, understanding that they are always contingent, and that tomorrow's experiments may prove everything they assumed previously to be suddenly incorrect. Merton's ethos demands, in short, open pursuit of scientific knowledge. Technology does not demand any of these. But because of the nexus between science and technology, pushed ever closer by various commercial and political forces, science is in danger of setting aside these important values. Technology, on the other hand, may very well have no ethos.

It is difficult to reasonably extend the ethos of science to technology, or for that matter, to conceive of anything like Merton's values in the pursuit of technology. People may well work in secret, applying the fruits of science to productive ends, they may well choose to work in groups, to obscure their labors, to devise means to defeat reverse-engineering. They certainly may be interested, in fame, profit, and world domination. I believe the ethics of any of these choices are independent of any technological ethos. But as a matter of pure self-interest, trying to extend any of these motivations or attitudes upstream toward the basic science will only hinder the ability of innovators to have access to new means to develop new technologies. As science and technology *merge*, which is arguably what is occurring with many converging technologies, attempts to hide, obscure, or protect the basic knowledge are not only unnecessarily expensive, but also self-defeating. Whatever IP future we choose, there will always be opportunities for fame and fortune of some kind from entering a marketplace with a killer app before others do. Shrewd inventors keep abreast of breaking science, and have the market sense to predict demands before even potential customers know they exist. The *application* of scientific discovery to something new and needed, or *wanted*, or creating a need or want, has made fortunes before, and always will, even absent IP. But taking advantage of the virtuous cycle between science and technology demands that the scientific ethos remain untainted by the perfectly ethical (but orthogonal) 'ethos' of the market (or lack thereof).

So why do converging technologies suggest that open innovation is preferable to traditional modes of innovation? Although the domains of science and technology do not fully overlap, the area of overlap is increasing. Moreover, we are discovering that openness in innovation is often a market advantage rather than a liability. Open innovation does not require one to divulge trade secrets, or to throw open wide one's entire stock of innovative creativity. But it does require rejecting IP. Market advantages can be gained more cheaply, and more creatively, than through the use of artificial monopolies. As automobile manufacturers discovered when they early on created 'patent pools' to prevent costly litigation, to standardize modes and parts, and to accelerate innovation in their industry, so too are innovators today learning that there is profit in sharing, and that competition is better than rent-seeking, or squatting on ideas that remain underdeveloped due to patents. Ethics does not demand open innovation in technologies, but it is becoming clear that without it the basic science is hampered, and with it, everyone benefits.

Converging technologies are shortening the distance between science and innovation, and decreasing the temporal gap between discovery and invention. The basic, foundational science is becoming more accessible, as with synthetic biology which can be practiced literally in garage-labs. The direction of home-built microfabrication as it is being pursued by serious hobbyists, using scientific techniques, should suggest that true MNT might actually be accomplished outside of formal labs as well. If the techniques and science behind self-replicating replicators is perfected in someone's garage, and not in a lab, this will be the final convergence not only of technologies, but also of science and technology. The missing bit, and the part that may well never be completed, is the holy grail of artificial intelligence (AI). This seems unlikely (so far) to converge as predicted with the technologies we have been discussing in this book. Its perfection, and its merger with nanowares, will pose a whole new set of ethical and practical issues. I have scrupulously avoided these issues in this book simply because AI seems to forever be just around the corner, but remains instead still so far away on a distant horizon. And this is, so far, fortunate.

How would we deal with AIs that innovate? When artifacts create artifacts, who could possibly 'own' the IP? When the *final* convergence occurs, we'll be thankful we have left IP far behind, or the robots will beat us all out of the marketplace. Terminators hell-bent on destroying humanity scare me much less than robots with patents. I'd rather be dead at the hands of runaway technology than have our own machines use our legal institutions to deprive us of our uniquely human gift of creativity.

A New Theory of IP and Its Role in Innovation

I have promised to develop a new theory of IP. In fact, it may not be new at all. Innovators in the 'useful' and 'aesthetic' arts worked for thousands of years outside the current IP paradigm, creating new things, profiting through their creativity, and enriching the sum of human achievement. The notion that ideas could be tied up, monopolized, and contained by laws is what is new. The theory of IP that I propose is one that has previously existed, and that began to be *revealed* through ICT, and will be completed with converging technology. It is simply this: all new man-made objects, intentionally produced, are *expressions* along a spectrum. That spectrum exists between primarily utilitarian and primarily 'aesthetic' uses. The ideas represented by these expressions belong to a 'commons by necessity,' which means that attempts to contain them are attempts to enclose an uncloseable space. These attempts are also attempts to curtail free expression and freedom of conscience, which ought to allow anyone to express any idea within the limits of the 'harm principle.'

The past 200 years of IP have been an interesting but doomed experiment. Borne out of liberalism, the seeds of its destruction are also inherent in liberalism. Anathema to free markets, contrary to freedom of expression, autonomy, and freedom of conscience, and also contrary to the scientific ethos, IP could only die eventually. But what replaces it? What was IP law *attempting* to accomplish? It was hardly an evil plot, but rather a misguided attempt to carry out the modern agenda of progress, human betterment, and freedom. But the objects of the law were mistaken. IP attempted to place limits on expressions in order to secure monopoly rights for innovators, to encourage their creativity, and to ultimately benefit the public. These ends could well have been achieved by focusing not on the objects, but rather the *actors*. Can we learn lessons about the functioning of rewards and incentives, and market-based commercial rights that can help ensure those incentives, by looking at other successful economic sectors?

The ethical considerations I have argued for above, regarding the duties of scientists and innovators, as well as freedoms of expression and conscience, apply equally to the consuming public. In any market exchange, there must be trust and reciprocity. Each party must perceive some benefit in the exchange, or the exchange is not likely to happen, unless it occurs under duress. Many markets thrive, as we have discussed above, without pre-existing monopolies granted by states. These markets prosper, secured only by the trust and expectation of benefit by all parties to market transactions. When you buy a loaf of bread, when you purchase a piece of clothing, when someone washes your car or watches your children, each party receives value, and that value is compensated on the one hand by the provision of not just a product, but a service. Someone else made that bread for you. Someone else baked that bread. You might have been able to do either of these, but the convenience of someone else performing these services was worth your payment. You could have washed your car, and you could well have watched your children, but other necessities made it worthwhile for you to *choose* to pay someone to do each of these for you, and to pay them (after the provision of the service) only because you likely felt they *ought* to be compensated for their service.

Creativity, as we have said before, will always be scarce. Not everyone will innovate, even as the tools that make innovation more accessible, in the form of nanowares, become ubiquitous and cheap. Creativity is a service. When someone introduces some new, man-made, intentionally produced object into the market, they are enriching our culture, increasing the store of human artifacts, and bettering lives (hopefully). Most people feel strongly that those who provide such a service ought to be compensated, just as bakers, farmers, grocers, babysitters, and car washers ought to be compensated. While many argue that the recent spate of peer-to-peer (P2P) technologies and the 'piracy' that they allow show that people cannot be trusted, and that states must create exclusive rights as well as enforce legal sanctions against 'thefts' of IP, I believe that the evidence shows that people will actually pay for value they receive, even if they know they do not need to.

Even while record companies bemoan the death of the music industry, the *independent* music industry is arguably thriving. As we saw above, there is a growing spate of examples of independent musicians adopting new means, utilizing the internet in various ways, to get attention for their work, and also to be compensated. P2P networks continue to thrive, and pretty much all content comes to be copied and shared on them. But the movies industry, television programs, and music have not collapsed. Perhaps P2P networks offer models by which these industries need to evolve. Given the ability to reach audiences, to distribute works, and to access the tools of production in media, the huge media companies of the past may be outmoded. Artists, filmmakers, and others, whose creativity was once hampered by the inaccessibility of the tools of production and distribution, can now be liberated by these forces. Musicians need have little more than a computer now to reach the world with their works. The future is wide open for those who wish to create, and there is no evidence that fears of IP 'theft' are keeping artists from creating. Nor does it appear that artists are suddenly starving because of the ability of the public to quickly share content through P2P networks. It is likely that profit margins would shrink if we suddenly ignored IP, and that business models would radically change. But the success of new markets, created in the wake of the success of underground P2P file sharing, suggests that creativity could in fact bloom.

One prominent example of how new markets can thrive is the iTunes store, and now the App Store, both introduced by Apple. Apple, which has historically been a strong IP proponent, and jealously guards its IP, has embraced, in my estimation, the vision of what the future of content really is. They are making money by bypassing the middlemen, the large record producers and content providers, and instead making new *markets* in which anyone could offer their content, take a larger cut of the costs as profits for themselves, and cut dependencies on old models of production and distribution. Even while Apple's IP policies are outmoded, what is critical is these new markets, and the freedom that creators have in accessing them, and in providing their content with or without traditional IP. Another promising element, particularly of the App Store, is that Apple has opened up its standards for app development so that anyone can make an app, and market it through the App Store. This has made a number of creative app developers wealthy overnight, and provided an expansive new venue for creativity and distribution. Because of the relative simplicity of app production, and new tools that streamline it, even those without particular expertise in coding have been able to access this new market.

Apple also produces tools that make music and film production (relatively) cheap and simple. These tools have revolutionized the ability to make professional film (film-quality video, really) and music at home, without large production studios, and with relatively little capital. Creativity is almost all the capital one needs to enter the market now with professional film and music. Over time, the prices of these tools continue to fall, and the learning curves for using these tools grow less steep. But above and beyond these tools, the genius of Apple is in creating the markets for the distribution of the products of creativity, and reducing barriers to entry for thousands of people who had good ideas, but previously could not, or would not, profit from them. Apple's failures to be totally open have provoked encouraging market reactions as well. Now that a US court has held them to be legal, markets for Apple-rejected apps are available through sites like Cydia, and these markets include paid apps as well. Let 10,000 app stores bloom.

These new markets seem to be thriving, even as the ability of consumers to 'pirate' new wares continues. One simple explanation for this is that people seem willing to pay US\$.99 for a song, US\$.99 for an app, or US\$9 for an album, because of the convenience of the iTunes and app stores, because these seem like more reasonable prices than the markups on traditional CDs, or maybe even because consumers think these prices are fair. It is worth compensating content creators for their services. Some people will not pay, and they might never have paid, but most people still seem willing to compensate those who make new things even where the threat of sanctions for not doing so is minimal, and the ability to take without paying is well within reach. Apple's IP policies are antiquated, and there should be more openness and transparency in its marketplaces, but their genius is in redesigning their business strategy to fit the future of content. They have embraced a vision of a future of content as a service, and not as products, and recognize the role of access to markets in facilitating creativity, both through providing new tools and in new marketplaces like iTunes and the App Store.

The Creativity Economy

This is the new theory of IP in a nutshell, and it is working. Creative production is a service, and while we cannot legitimately enclose ideas, we still *ought* to compensate people for providing services that we find valuable. The market should dictate the cost of those services, rather than allowing state-sanctioned monopolies to skew prices. When faced with new creative content that is valued, consumers will pay for that value, not out of duress, but based upon trust, as other market transactions for services have long worked by trust.

Other examples of the emergence of new market models abound, and as with Apple, their IP policies are not always consistent, but they demonstrate the fact that given new markets, people will devise new and creative ways to profit from their creativity, with or without IP. Google's platforms for services are open, for the most part, as are Facebook's. Even while each obscures a fair amount of their operations, they have each created new marketplaces that open up content provision to those who might never have been able to realize their creativity before. Facebook apps are created by large conglomerates and basement programmers, and available to all Facebook users alike. As with the App Store, Facebook's app platform standards are available to everyone who wants to try to develop an app. Although Google keeps its most valuable asset (its algorithms for its search engine) as a trade secret, the open standards it employs for its service platforms allow the development of new apps by anyone, and the provision of profitable services on top of those platforms by new market players.

Each of these imperfect examples illustrates the model that nanowares will likely follow, and a mechanism by which openness of standards and the foundational science can help to facilitate profits by a new creative class. Physical goods, produced using the methods of nanowares, will be the apps of the future. Those who are embracing openness in the foundational science and technology are providing the open platform upon which artifactual 'apps' will be designed and distributed in the nanoware future. As with the App Store, iTunes, or any of the examples discussed above, people will pay for these artifactual 'apps' because they value the service of those who design them, the costs will likely be reasonable and fair, and people will be willing to compensate (as they are now, despite P2P networks) for the convenience of downloading something they want or need. Nanoware will be the ultimate killer app, enabling people to fully realize their creativity, to enter marketplaces without friction, and to quickly access any tool they need or desire at little cost. Even as the costs of nanowares may fall toward zero, the margin of profit will be comparatively high as the tools of production become cheap and ubiquitous, and the expenses associated with market entry and distribution disappear.

Creativity will be the new capital, and those who have it will continue to innovate and profit, and consumers will be able to profit from the products of creativity. Nanowares, as we have envisioned them here, and as those currently working to see their development envision them, will involve the final unleashing of the possibilities of human creativity, the emergence of expansive new markets for the fruits of that creativity, and the possibility of market entry for anyone with nothing more than an idea. Trust will be the basis of these new markets, even as it is the basis of markets for bread and other goods, provided as services, since the beginning of commerce.

Nanowares and Converging Philosophical Inquiries

I have addressed only superficially a number of issues consuming philosophers with the introduction of nanowares, primarily in the areas of ethics of risk, safety, and security. In so doing, I have argued that nanowares do not present wholly unique concerns, although they are certainly serious. Converging technologies require us to extend our current philosophical foci on risks, and on the natures of technical artifacts, but they may not require us to rethink anything fundamentally. Risks, duties, and even utilitarian calculi can be adopted to deal with challenges posed by converging technologies, and these will remain fruitful realms of inquiry. But perhaps the balkanization of applied ethics can be resolved through nanowares, even as the technologies themselves converge. The emergence of medical ethics, bioethics, and other sub-categories of applied ethics may ultimately also converge again with nanowares. Synthetic biology is, after all, an engineering paradigm, applied to traditionally 'biological' substrates. Does the substrate make all the difference, or are the duties and rights involved the same regardless of the applied field? I have argued above that the duties owed by scientists and engineers are in fact the same, regardless of the medium. Nanowares are an opportunity for us to reconsider the canard that 'the medium is the message,' and instead to say: the medium doesn't matter at all.

Do we need a philosophy of technology, or of aesthetics, or an applied ethics in medicine, and another in nanotechnology, or do the underlying objects and principles, as they converge through nanowares, reveal that these subfields are all intimately related. I would argue that philosophers of technology and philosophers of art unfairly distinguish among the objects of their field, just as the law of IP has illogically distinguished between utilitarian and aesthetic objects. All artifacts have uses, so why do we feel it necessary to treat aesthetic uses and other uses ('utilitarian uses' would be redundant) as distinct? Perhaps there are interesting cultural or psychological reasons why we make these distinctions, and we ought to examine the reasons we make them (particularly if we discover that other cultures or certain people do not make such distinctions), but perhaps nanowares suggest that various fields of philosophical study ought to converge as well, or at least critically examine the reasons for their divergence.

Physics, chemistry, biology, engineering, design, and a host of other disciplines are implicated in nanowares. This offers an exciting realm for reconsidering the nature of our sciences, technologies, and artifacts, and the relationships among cultures, individuals, and their products. I have only touched on one small area of particular concern to me, namely IP, and some ethical considerations. But there is certainly a great deal of interesting analysis that remains to be done even as nanowares continue to develop. Each new technical field offers philosophers new opportunities for inquiry. This is one reason I have focused my work on the philosophy of science and technology. We should be open and interested in what is new. But even while the objects of our study may be new, we should be wary of multiplying principles, or finding uniqueness where there is none.

When ICT emerged as an area of philosophical inquiry, a number of people suggested that cyberspace was somehow fundamentally different from other 'spaces.' One typical response to the problem of software's dual status as both patentable and copyrightable was that it was somehow a new 'hybrid' object of a type that had never before existed. I did not think so when that was proposed, and I have grown more certain that it is not in any way unique. It is an expression of the same kind as *Ulysses* or as a steam engine, and the distinctions we previously drew among novels and engines were never firmly grounded. We could well have treated all of the objects of IP as I suggested, with something like copyright, only with shorter terms. But something else happened that I did not anticipate. The market found solutions, and producers began to skirt existing IP norms consciously, choosing open source solutions, rooted in private contract. Private contract is itself founded upon ancient principles. It has proven to be the working foundation of most market phenomena, and as long as it remains backed up by institutions by which disagreements can be resolved, it should be preferred over artificial monopolies. The emergence and success of open source, copyleft, creative commons, and other private means of promoting and profiting by innovation without IP serve as models for the nanoware future. It is not surprising that so many who are already engaged in building that future have adopted these models, or models like them. Looking back at their emergence, it is clear that my philosophical concerns were well founded, but my suggestions for resolving them unnecessarily conservative. The market figured it out, and better models emerged. I am confident now that the trends that are already extending these models to nanowares will succeed, and that like it or not, nanowares embody the end of IP. It's about time.

Case Study

The Final Convergence: From Idea to Market

Nanowares are here, and they are evolving. In each of their various permutations, we will have choices about how we choose to move ideas into markets. Throughout this text, we have considered the practical, philosophical, and even ethical implications of choosing to treat our artifacts in particular ways. Now let's examine how various forms of nanowares might be placed into the stream of commerce under several different scenarios.

The Nano-now: A RepRap Designed Object

Using off the shelf CAD/CAM design software, one can design and prototype objects using the RepRap Mendel. Suppose you design a salt-and-pepper shaker set, which could be easily constructed on a Mendel, using the extruded plastics, and a CAD/CAM designed form. Let's assume that it is unique, incorporating design elements that distinguish your salt-and-pepper shakers from all others on the market, either functionally, or aesthetically, or both. Under current patent law, only truly new, non-obvious, and useful elements of your shakers could be patented, but let's assume also that you have developed such elements. Arguably, if you choose to patent your shakers, you will be forced into capitalizing the creation of a robust production and distribution plan, or license the idea to others. Using the current patent system, you could not patent the design

Case Study: *The Final Convergence: From Idea to Market* (continued)

specifications themselves, though you could copyright them. If you want to distribute your shakers via a nanowares infrastructure, as designs that RepRap owners could use to reproduce the shakers themselves, you would need to copyright your designs, patent the final object, and sell the designs with a license of some sort that entitles purchasers to reproduce the patented shakers for their own use. This would enable you to enter the market with your product and only pay for the patent process, utilizing the nanowares infrastructure, eliminating yourself from the burden of raising capital for production and distribution, and helping to protect your idea from 'pirating.' Could such a process work?

There is no reason to think that consumers won't pay monopoly rents for items they value, and end user license agreements are used now for the distribution of software, parts of which may be both copyrighted and patented. While clunky, this option exists as a means of preserving strong IP protection, using the new and valuable processes involved with won't emerging nanowares, and allowing those who create new things to enter the market perhaps a bit more easily than in the past. The drawbacks to this approach are the same as in software, in that end users often agree to the terms in order to secure the goods, and then violate the terms as they please. Pirated versions of software proliferate on P2P networks, none of which would be available except for some user's violation of an end user licensing agreement (EULA). You could ensure that some people abide by the license, you would get some of the proceeds of the sales of your product, but there is no way of stopping those who decide to distribute your works despite the EULA. Once they do decide to violate the EULA, and your designs end up being distributed without your awareness, anyone with a RepRap will be able to duplicate your shakers. If all they want is a 'free' salt-and-pepper shaker set, then you will have virtually no way of stopping this, nor of seeking out those who possess illicit copies, short of some sort of technological means that would provide a means to track them down.

These are pragmatic obstacles. We have not discussed at all other reasons to avoid the strong IP route. Supporters of IP may well choose this route, and they may be able to succeed with new and inventive objects that are sufficiently desired, as long as their products are priced within a range that the market will tolerate. After all, as we have discussed above, IP is working for new markets like iTunes, the App Store, and even the Cydia store. People will pay (reduced or reasonable) monopoly prices either out of some sense of duty, desert, or merely due to convenience. One advantage of the emerging nanowares infrastructure, even for those who choose to abide by a strong IP regime, is that the costs of capitalizing new products are significantly reduced. In fact, in this scenario, besides the labor and creativity involved in dreaming up a great new salt-and-pepper shaker, the major costs would be involved in securing a patent. This option uses existing institutions and models, and among the only reasons to not pursue it would be the pragmatic inability to police distribution of your design, some sense that strong IP will not be fair, efficient, or worthwhile, etc. Perhaps you even feel that you could profit, and innovation in general would benefit, from some other institutional approach.

Another reasonable option might be to copyright the entire object, and avoid the costs associated with patenting. This approach makes metaphysical sense, as described at length throughout this text. The design and the final product would each be considered copyrighted expressions, and each would be susceptible to the same protections and monopoly benefits afforded unique expressions. While this option makes metaphysical sense, it is not yet supported by any institutional framework. Current laws do not consider *non-utilitarian* expressions to be on par with utilitarian ones. We might well hope that some day the law catches up to the facts of the matter, but if one were to try this now, it would have to be done with the knowledge and fortitude to pursue legal reform through a test case.

Case Study: *The Final Convergence: From Idea to Market* (continued)

Finally, you might choose to distribute your salt-and-pepper shakers under some sort of 'copyleft,' creative commons, or similar license, or no license at all. As with the previous option, the costs for market entry are essentially nothing (excluding the costs of labor and creativity). So how would you get paid? What reward would you receive for your creativity? The rewards might certainly be monetarily lower than you *might* receive under strong IP, assuming there were sufficient market demand for your product (no product is guaranteed to be desired) and sufficient consumers willing to abide by a EULA. But this does not mean you might not do even *better*. It is conceivable that consumers will so desire your product that, as with innovative means of rewarding artists through donations, or payment for services, you might well be paid by consumers of your product according to their valuing of your creativity. A 'donate' button by the file might help to urge those who appreciate your creation to reward you for your efforts. If you wanted to help encourage this, you might well offer the design as a form of 'shareware,' which although it could be used without payment, might prod users to pay you according to their valuation of your services as a creator.

This option requires no institutional adjustment, and offers some benefits as well as apparent drawbacks over strong IP or a unified copyright regime. This approach operates merely on *trust*. It assumes that people will pay for services they find valuable. It also requires accepting that some people do not pay for valuable services, even when they benefit. This is true for most other sectors of the economy, so why should we not accept it for creative sectors? Creators too may benefit under such a scheme, as their creations will be prone to improvement, and the ever-increasing stock of innovative and improving goods will grow, presumably. But this final, perhaps, utopian approach also justifies itself based upon the expectation that the need for capitalizing the development, production, and distribution of new products falls to essentially zero. As a creator, only your time, intellectual effort, and creativity involved in developing the revolutionary salt-and-pepper shaker are an expense. In the past, you would need to invest in prototyping, mass producing, and distributing new products, and no IP could ever guarantee market success as no inventor or author can ever truly predict the desires of consumers. This is why most new products fail, and how fortunes have sometimes been lost in the process even of bringing revolutionary products to the market. In this scenario, though, what has been lost is the time, and perhaps the 'labor' of creation. Success, we hope, will still be rewarded, and the marketplace will grow as new efforts at creation can potentially earn market share based solely on the desires of consumers, without the costly burdens of failed products.

Each of these scenarios fits just as well with the far nano-future in which MNT allows something like Stephenson's 'matter compliers' to replicate any product, anywhere. What applies to RepRap-produced salt-and-pepper shakers applies to any nano-based object of an MNT future. These are all choices, and those who are paving the way for the nano-future can pursue them as they see fit. I have tried to make the case that the latter path is most virtuous, provides the best future for innovation itself, and is justified by the economics of nanowares. Now let's see where the technology takes us.

Notes

Preface

1   My ongoing commentary about the case can be accessed online at http://whoownsyou-drkoepsell.blogspot.com/

Introduction

1   As I began to discuss in *The Ontology of Cyberspace: Law, Philosophy, and the Future of Intellectual Property*, Chicago, IL: Open Court, 2000.

2   Or more broadly 'nanowares' which encompasses a range of technologies between full MNT and desktop 3D printing, as I will discuss in more depth soon.

3   See generally, Dipert, R. (1993), *Artifacts, Art Works, and Agency*, Philadelphia, PA: Temple University Press.

4   See generally, Williams, T. and Singer, C. J. (1954), *A History of Technology*, vols. 1–6, Oxford: Clarendon Press; Bronowski, J. (1973), *The Ascent of Man*, Toronto, ON: Little, Brown and Company.

5   Daniels, P. and Bright, W. (1996), *The World's Writing Systems*, Oxford: Oxford University Press, p. 146.

6   See generally, Smith, B. D. (1995), *The Emergence of Agriculture*, Scientific American Library.

7   Braudel, F. (1992), *Civilization and Capitalism: Vol. 2, The Wheels of Commerce* (trans. S. Reynolds), Berkeley, CA: University of California Press.

8   Wolf, A. (1950), *A History of Science, Technology, and Philosophy*, vol. I, New York: Harper Torchbooks, pp. 54–70.

9   See generally, Morton, D. L. Jr. and Gabriel, J. (2007), *Electronics: The Life Story of a Technology*, Maryland: The Johns Hopkins University Press.

10   Sobel, D. (1996), *Longitude: The True Story of a Lone Genius Who Solved the Greatest Scientific Problem of His Time*, London: Penguin Books.

11   http://en.wikipedia.org/wiki/Mallock_machine [accessed 22 September 2010].

12   Morton and Gabriel, *Electronics* (and now, materials other than silicon as well).

13   Campbell-Kelly, M. and Aspray, W. (2004), *Computer: A History of the Information Machine*, Boulder, CO: Westview Press.

14   Campbell-Kelly and Aspray, *Computer*.

15   I describe and defend this notion of the Jacquard Loom in *The Ontology of Cyberspace*.

16   209 US 1 (1908).

17   Pub. L. 94-553 (19 October 1976); 9 0 Stat. 2541.

18   As I argued we should do in *The Ontology of Cyberspace*.

19   I develop this argument at length in both *The Ontology of Cyberspace*, and in *Who Owns You*? If you want to follow this argument without buying those books, you can read more about it on my blog for *Who Owns You*?. Available at http://whoownsyou-drkoepsell.blogspot.com/

Chapter One

1 This chapter includes material from, and expands upon my article (2009),
‘Let's Get Small: An Introduction to Transitional Issues in Nanotech and
Intellectual Property’, *Nanoethics*, 3(2): 157–66 which is published under an
Open Access license: DOI citation: 10.1007/s11569-009-0068-9.

2 Published first in *Engineering and Science* (1960), republished at http://www.
zyvex.com/nanotech/feynman.html [accessed 22 September 2010].

3 Drexler, E. (1986), *Engines of Creation: The Coming Era of Nanotechnology*,
New York: Anchor Books.

4 Vinge, V. (1993), ‘Technological Singularity’ paper presented at the Vision 21
Symposium, expanding on a concept developed in his short story ‘Marooned in
Realtime’ as well as his article ‘First Word’ in *OMNI* magazine, 1983.

5 Kurzweil, R. (2000), *The Age of Spiritual Machines*, London: Penguin Books.

6 Joy, B. (2000), ‘Why the Future Doesn't Need Us’, *Wired*, 8(4), April.

7 Kuhn, T. (1962), *The Structure of Scientific Revolutions*, Chicago, IL: University
of Chicago Press.

8 During a talk in 1977 for the World Future Society, in Boston. http://www.snopes.
com/quotes/kenolsen.asp [accessed 9 September 2010].

9 See, for example: Ach, J. S. and Siep, L. (eds) (2006), *Nano-bio-ethics: Ethical
Dimensions of Nanobiotechnology*; Cameron, N. M. de S. and Mitchell, M. E. (eds)
(2007), *Nanoscale: Issues and Perspectives for the Nano Century*; Berne, R. W.
(2006), *Nanotalk: Conversations with Scientists and Engineers about Ethics,
Meaning, and Belief in the Development of Nanotechnology*; Allhoff, F. and
Lin, P. (eds) (2009), *Nanotechnology and Society: Current and Emerging
Ethical Issues*; Schummer, J. and Baird, D. (eds) (2006), *Nanotechnology
Challenges: Implications for Philosophy, Ethics and Society*; Roco, M. C.
and Bainbridge, W. S. (eds) (2006), *Nanotechnology: Societal Implications*; Miller,
J. C., Serrato, R., Represas-Cardenas, J. M., and Kundahl, G. A. (2004),
*The Handbook of Nanotechnology: Business, Policy, and Intellectual Property
Law*; Boucher, P. M. (2008), *Nanotechnology: Legal Aspects*.

10 Pelley, J. and Saner, M. (2008), ‘International Approaches to the Regulatory
Governance of Nanotechnology’, RGI Report, School of Public Policy and
Administration at Carleton University in Canada.

11 Seed, P. (2001), ‘Taking Possession and Reading Texts: Establishing the Authority
of Overseas Empires’, in S. N. Katz, J. M. Murrin, and D. Greeberg (eds), *Colonial
America: Essays in Politics and Social Developments*, 5th edn, New York:
McGraw-Hill, pp. 22, 26.

12 Cheung, S. N. S., ‘Property Rights in Trade Secrets’, *Economic Inquiry*, 20(1): 40.

13 Chicago, IL: Open Court (2000).

14 http://fab.cba.mit.edu/about/faq/ [accessed 10 June 2009].

15 http://fabathome.org/wiki/index.php?title=Fab%40Home:Overview [accessed
10 June 2009].

16 http://reprap.org/bin/view/Main/WebHome [accessed 10 June 2009].

17 25 November 2006, http://www.guardian.co.uk/science/2006/nov/25/
frontpagenews.christmas2006 [accessed 10 June 2009].

18 25 November 2006, http://www.guardian.co.uk/science/2006/nov/25/
    frontpagenews.christmas2006 [accessed 10 June 2009].

19 See, *Who Owns You? The Corporate Gold Rush to Patent Your Genes*.

Chapter Two

1 http://www.iter.org [accessed 22 September 2010].

2 Drexler, E. (1992), *Nanosystems: Molecular Machinery, Manufacturing, and
   Computation*, New York: John Wiley & Sons.

3 Stephenson, N. (2000), *The Diamond Age*, New York: Bantam Books.

4 Drexler-Smalley Debates (original text) http://pubs.acs.org/cen/coverstory/8148/
   8148counterpoint.html [accessed 22 September 2010].

5 Smalley, R. E. (2001), 'Of Chemistry, Love, and Nanobots', *Scientific American*,
   285: 76–7.

6 Drexler, E., Forrest, D., Freitas, R. A. Jr., Hall, J. S., Jacobstein, N., McKendree, T.
   and Merkle, R. (2001), *On Physics, Fundamentals, and Nanorobots*; Peterson, C.
   (2001), *Rebuttal to Smalley's Assertion that Self-replicating Mechanical
   Nanorobots Are Simply Not Possible*, Institute for Molecular Manufacturing.

7 http://www.softmachines.org/wordpress/?p=175 [accessed 22 September 2010];
   Jones, R. A. L. (2008), 'Rupturing the Nanotech Rapture', *IEEE Spectrum*'s Special
   Report: The Singularity.

8 Attempts to manipulate biological systems and parts to conduct, essentially,
   MNT. This is inspired by the fact that many biological phenomena are essentially
   nanoscale phenomena.

9 http://www.nanotechproject.org/ [accessed 22 September 2010].

10 Koepsell (2000), *The Ontology of Cyberspace*.

11 Koepsell (2009), *Who Owns You?*

12 Watt, S. (1997), 'Introduction: The Theory of Forms (Books 5–7)', Plato: *Republic*,
    London: Wordsworth Editions, pp. xiv–vi.

13 And arguably a third sort of thing we might call 'accidents'.

14 See, *The Ontology of Cyberspace*.

15 Simons, P. and Melia, J. (2000), 'Continuants and Occurrents', *Proceedings of the
    Aristotelian Society*, Supplementary Volume 74: 59–75, 77–92; see Munn, K. and
    Smith, B. (2008), *Applied Ontology: An Introduction*, p. 268.

16 For more on Basic Formal Ontology, see http://www.ifomis.org/bfo/1.1
    [accessed 22 September 2010].

17 This is an explicitly Aristotelean view of universals, and contrasts with Plato's
    metaphysics. *Metaphysics*, Z 13; *Posterior Analytics*, I 5, 74a: 25–32.

18 A method of placing an order for an item comprising: under control of a client
    system, displaying information identifying the item; and in response to only a
    single action being performed, sending a request to order the item along with
    an identifier of a purchaser of the item to a server system; under control of a
    single-action ordering component of the server system, receiving the request;
    retrieving additional information previously stored for the purchaser identified
    by the identifier in the received request; and generating an order to purchase the

requested item for the purchaser identified by the identifier in the received request using the retrieved additional information; and fulfilling the generated order to complete purchase of the item whereby the item is ordered without using a shopping cart ordering model.

Chapter Three

1 http://educhoices.org/articles/How_to_Make_Almost_Anything_OpenCourse Ware_MITs_Free_Graduate_Level_Course_on_Fabrication_and_Design.html [accessed 22 September 2010].

2 Gershenfeld, N. (2005), *FAB: The Coming Revolution on Your Desktop – From Personal Computers to Personal Fabrication*, New York: Basic Books.

3 http://fab.cba.mit.edu/about/charter/ [accessed 22 September 2010].

4 http://www.fabathome.org/wiki/index.php/Fab%40Home:_Model_2_Overview [accessed 22 September 2010].

5 http://reprap.org/wiki/Mendel [accessed 22 September 2010].

6 'There are around 1.5 million patents in effect and in force in this country, and of those, maybe 3,000 are commercially viable' (Richard Maulsby, director of the Office of Public Affairs for the US Patent & Trademark Office, says). 'It's a very small percentage of patents that actually turn into products that make money for people' (Richard Maulsby quoted in Klein, K. E. (2005), 'Smart Answers, "Avoiding the Inventor's Lament"', *Business Week*, 10 November).

7 Madden, M. (2009), 'The State of Music Online: Ten Years after Napster', *Pew Internet Reports*, 15 June. Available at http://www.pewinternet.org/Reports/ 2009/9-The-State-of-Music-Online-Ten-Years-After-Napster/The-State-of-Music-Online-Ten-Years-After-Napster.aspx?view=all [accessed 22 September 2010].

8 Welsh, C. A. (1948), 'Patents and Competition in the Automobile Industry', *Law and Contemporary Problems*, 13(2), The Patent System (Spring): 260–77.

9 Bill Gates admits his proficiency in coding grew from thousands of hours of free time on the computer at his schools in which he could muck about and write code to his heart's content. At that stage, he had not begun to use IP to protect his code, and the code of others on the system with whom he collaborated, learned, and honed his skills was similarly unprotected. See, Gladwell, M. (2008), *Outliers*, London: Little, Brown and Company.

10 'The History of the GPL'. Available at http://www.free-soft.org/gpl_history/ [accessed 22 September 2010].

11 See http://en.wikipedia.org/wiki/List_of_free_and_open_source_software_ packages [accessed 22 September 2010].

12 Thurston, R. (2006), 'HP: Open Source Can be More Profitable than Proprietary'. Available at http://www.ZDNet.co.uk [accessed 29 October 2010]; http://www. zdnet.co.uk/news/application-development/2006/10/29/hp-open-source-can-be-more-profitable-than-proprietary-39284344/ [accessed 22 September 2010].

13 http://seekingalpha.com/article/10166-chart-software-companies-gross-profit-margins [accessed 22 September 2010].

Chapter Four

1 This principle, which puts the burden of proof of safety upon those who would pursue something (such as a technology) given the possibility of some harm, is statutorily required in the European Union. Recuerda, M. A. (2006), 'Risk and Reason in the European Union Law', *European Food and Feed Law Review*, 5.

2 Taebi, B. (2010), 'Sustainable Energy and the Controversial Case of Nuclear Power', in R. Raffaelle, W. Robison and E. Selinger (eds), *Sustainability Ethics: 5 Questions*, Copenhagen: Automatic Press.

3 Before law school, I had the honor of serving as a clerk for New York State attorneys who sued Hooker Chemical on behalf of the State for damages arising from the massive dumping of chemical byproducts at Love Canal, in Niagara Falls, NY. That case was but one of many that uncovered what many allege were derelictions of duties owed by chemical companies to residents, both present and future, of areas where chemicals were made and wastes disposed, without the sort of precautions now required. See Glaberson, W. (1990), 'Love Canal: Suit Focuses on Records from 1940's', *The New York Times*, 22 October.

4 The Belmont Report is a significant statement of ethical principles pertaining to research on human subjects in the United States. 'The Belmont Report: Ethical Principles and Guidelines for the Protection of Human Subjects of Research' by The National Commission for the Protection of Human Subjects of Biomedical and Behavioral Research, 18 April 1979. This report is now the foundation upon which rules and guidelines for determining the ethics of studies on humans are based. Available at http://ohsr.od.nih.gov/guidelines/belmont.html [accessed 22 September 2010].

5 Milgram, S. (1963), 'Behavioral Study of Obedience', *Journal of Abnormal and Social Psychology*, 67: 371–8. DOI: 10.1037/h0040525; Milgram, S. (1973), 'The Perils of Obedience', *Harper's Magazine*, December, pp. 62–77.

6 Tuskeegee Study – Timeline. NCHHSTP. Centers for Disease Control 2008-06-25. Available at http://www.cdc.gov/tuskegee/timeline.htm [accessed 22 September 2010].

7 Koepsell (2009), 'On Genies and Bottles: Scientists' Moral Responsibility and Dangerous Technology R&D', *Science and Engineering Ethics*, 16(1): 119–33 (2010 – in print). Online first/Open Access DOI citation: 10.1007/s11948-009-9158-x.

8 http://www.foresight.org/guidelines/current.html [accessed 22 September 2010].

9 http://ec.europa.eu/nanotechnology/pdf/nanocode-rec_pe0894c_en.pdf [accessed 22 September 2010].

10 The Avalon Project – The Laws of War, Yale Law School, Lillian Goldman Library. Available at http://avalon.law.yale.edu/subject_menus/lawwar.asp [accessed 22 September 2010].

11 Press Briefing by Homeland Security Director Tom Ridge, Health and Human Services Secretary Tommy Thompson, CDC Emergency Environmental Services Director, Dr. Pat Meehan 29 October 2001. Available at http://www.presidency.ucsb.edu/ws/index.php?pid=79187 [accessed 22 September 2010].

12 Cello, J., Paul, A. V., and Wimmer, E. (2002), 'Chemical Synthesis of Poliovirus cDNA: Generation of Infectious Virus in the Absence of Natural Template', *Science*, 297(5583): 1016–8. DOI:10.1126/science.1072266. PMID 12114528.

13 http://gspp.berkeley.edu/iths/UC%20White%20Paper.pdf [accessed 22 September 2010].

14 http://synbiosafe.eu; http://www.idialog.eu/uploads/file/Synbiosafe-Biosecurity_ awareness_in_Europe_Kelle.pdf [accessed 22 September 2010].

15 http://ec.europa.eu/european_group_ethics/docs/opinion25_en.pdf [accessed 22 September 2010].

16 http://synthethics.eu/ [accessed 22 September 2010].

17 Garfinkel, M. S., Endy, D., Epstein, G. L., and Friedman, R. M. (2007), *Biosecurity and Bioterrorism: Biodefense Strategy, Practice, and Science*, 5(4): 359–62. DOI:10.1089/bsp.2007.0923.

Chapter Five

1 This chapter incorporates and expands upon my article, 'Things in Themselves: Redefining Intellectual Property in the Nano-age', *Journal of Information Ethics*, 21(1) (Spring 2011), and is used with permission.

2 'To promote the progress of science and the useful arts, by securing for limited times to authors and inventors the exclusive right to their respective writing and discoveries'; enacted through 35 US Code sec. 1-376, The US Patent Act.

3 Machlup, F. and Penrose, E. (1950), 'The Patent Controversy in the Nineteenth Century', *The Journal of Economic History*, 10(1): 1–29.

4 35 US Code sec. 154(a)(2).

5 See, *Gottschalk v. Benson*, 409 US 63 (1972), in which the US Supreme Court denied a patent to an algorithm that was essentially a mathematical formula, arguing that patenting an abstract idea or law of nature would not suit the purposes of patent.

6 See *Diamond v. Diehr*, 450 US 175 (1981), in which the US Supreme Court allowed a patent utilizing a well-known mathematical algorithm. Although they did not explicitly overrule Gottschalk, they reasoned that the algorithm when introduced into the machinery of a computer renders the previously unpatentable algorithm suddenly patentable, as part of a 'whole' invention.

7 Heim, M. (1993), *The Metaphysics of Virtual Reality*, New York, NY: Oxford University Press, pp. 81–3; see also, Benedikt, M. (1991), *Cyberspace: First Steps* Cambridge, MA: MIT Press.

8 McLuhan, M. and Fiore, Q. (1967), *The Medium is the Massage* [*sic.*], London: Penguin Books. McLuhan's son has clarified this strange title subsequently, '[...]the title was a mistake. When the book came back from the typesetter, it had on the cover "Massage" as it still does. The title should have read The Medium is the Message but the typesetter had made an error. When Marshall McLuhan saw the typo he exclaimed, "Leave it alone! It's great, and right on target!" Now there are four possible readings for the last word of the title, all of them accurate: "Message" and "Mess Age", "Massage" and "Mass Age"'. Eric McLuhan in: Baines, P., *Penguin by Design, A Cover Story 1935–2005*, Penguin Books, p. 144.

9 Koepsell (2000), *The Ontology of Cyberspace* arguing that cyberspace is nothing particularly new, and that all its components are expressions of known, material types.

10 See, for example, Stallman, R. (2006), *GPLv3, Presentation*, 25 February, Brussels, BE. Available at http://sweden.fsfe.org/projects/gplv3/ [accessed 22 September 2010].

Chapter Six

1 This chapter expands upon and incorporates my article (2010) 'Authorship and Artifacts: Remaking IP Law for Future Objects', *The Monist*, 93(3): 481–92 and is used by permission.

2 Coryton, J. (1855), *The Law of Letters-patent*, London: H. Sweet, pp. 4–5.

3 Coryton, *The Law of Letters-patent*, pp. 17–21, 131.

4 See generally, Koepsell (2000), *The Ontology of Cyberspace*.

5 See, Koepsell (2009), *Who Owns You?*

6 See, Coryton (1855), p. 88.

7 *Gottschalk v. Benson*, 409 US 63 (1972).

8 *Parker v. Flook*, 437 US 584 (1978).

9 *Diamond v. Diehr*, 450 US 175 (1981).

10 John Searle divides reality into the world of 'brute facts', which exists with or without human intentions, and the world of 'social reality', which is composed of the institutions, predicated upon brute facts, depending upon them in part, but founded ultimately upon human intentionality. Searle, J. R. (1997), *The Construction of Social Reality*, New York: Free Press.

11 Although in copyright, if the alleged infringer was innocent, and did not know of the existence of the other work, this is a defense. There is no such defense in patent.

Chapter Seven

1 In other words, positive (human made) law cannot contradict natural law. Consider the law of property. I argue that the legal institution of ownership is just because it is based upon the brute fact (the pre-institutional fact) of possession. Laws of ownership are just when they properly uphold brute facts of possession. The old canard that 'possession in 9/10ths of the law' is true, and reflects the fact that the law treats possession as *a priori* justification for continued possession, absent some other valid claim to prior possession, and some evidence of *unjust* dispossession. In legal institutions regarding *real* property (which is exclusive and rivalrous, meaning no two persons can simultaneously possess it), we have created a system of registration to track the proper, just possessors and thus owners. Based on all this, if some regime wished to declare *all* property to be *theft*, for instance, that regime would be unjust since it violates grounded (natural) rights of possession and just legal institutions of ownership. This is a more or less Lockean theory of natural rights to property, with a bit of an Austrian twist.

2 Boldrin and Levine, p. 2.

3 Menger, C. (1963), *Problems of Economics and Sociology* (trans. F. J. Nock, ed. L. Schneider), Urbana: University of Illinois Press, pp. 213–15.

4 Locke, J. (1988), *Two Treatises of Government* (ed. P. Laslett), Cambridge: Cambridge University Press. See, for example, 2d Treatise, Sections 171, 220.

5 Rosen, W. (2010), *The Most Powerful Idea in the World*, New York: Random House.

6 Machlup and Penrose (1954).

7 Moser, P. (2005), 'How Do Patent Laws Influence Innovation? Evidence from Nineteenth-century World's Fair', *American Economic Review*, 95(4): 1214–36.

8 Williams, H. (2009), 'Intellectual Property Rights and Innovation: Evidence from the Human Genome', 30 December, unpublished manuscript. Available at http://www.nber.org/~heidiw/papers/5_12_10a_hlw.pdf [accessed 22 September 2010].

9 35 USC § 200–212[1], and implemented by 37 C.F.R. 401.

10 See generally, Greenberg, D. (2007), *Science for Sale: The Perils, Rewards, and Delusions of Campus Capitalism*, Chicago, IL: University of Chicago Press.

11 Austin, J. (1832), *The Province of Jurisprudence Determined* (ed. W. Rumble), Cambridge: Cambridge University Press.

12 Kelsen, H. (1967), *Pure Theory of Law* (trans. M. Knight), Berkeley, CA: University of California Press.

13 Hart, H. L. A. (1958), 'Positivism and the Separation of Law and Morals', *Harvard Law Review*, 71: 593 repr. in his (1983) *Essays in Jurisprudence and Philosophy*, Oxford: Clarendon Press.

14 See generally, Rawls, J. (1999), *A Theory of Justice*, revised edn, Cambridge, MA: Harvard University Press; Herman, B. (ed.) (1999), *Lectures on the History of Moral Philosophy*, Cambridge, MA: Harvard University Press.

15 Reinach, A. (1913), 'The *A Priori* Foundations of Civil Law' (trans. J. Crosby), *Aletheia*, 3(1983): 1–142.

16 Dworkin, R. (1977), *Taking Rights Seriously*, London: Duckworth.

17 Many thanks to Shane Wagman for providing these examples.

18 Kot, G. (2009), *Ripped: How the Wired Generation Revolutionized Music*, Scribner, p. 215.

19 http://www.nytimes.com/2010/02/20/opinion/20kulash.html [accessed 22 September 2010].

20 http://nymag.com/daily/entertainment/2010/03/new_ok_go_video.html [accessed 22 September 2010].

21 http://2007.sxsw.com/music/showcases/band/26755.html [accessed 13 June 2009].

22 http://ghosts.nin.com/main/order_options [accessed 22 September 2010].

23 http://ghosts.nin.com/main/order_options [accessed 22 September 2010].

24 Kot, p. 249.

25 http://www.jillsnextrecord.com/ [accessed 22 September 2010].

26 Thompson, K., 'Future of Music Coalition', *New Business Models* (on file with the author).

Chapter Eight

1 This chapter includes and expands upon my article (2010) 'On Genies and Bottles: Scientists' Moral Responsibility and Dangerous Technology R&D', *Science and Engineering Ethics*, 16(1): 119–33 (in print). Online first/Open Access DOI citation: 10.1007/s11948-009-9158-x.

2 Atlas, R. M. and Dando, M. (2006), 'The Dual-use Dilemma for the Life Sciences: Perspectives, Conundrums, and Global Solutions', *Biosecurity and Bioterrorism: Biodefense Strategy, Practice, and Science*, 4(3): 276–86.

3 Preston, R. (2003), *The Demon in the Freezer*, New York: Fawcett.

4 Preston, *The Demon in the Freezer*.

5 Preston, *The Demon in the Freezer*.

6 Cohen, H. W., Gould, R. M., and Sidel, V. W. (2004), 'The Pitfalls of Bioterrorism Preparedness: The Anthrax and Smallpox Experiences', *American Journal of Public Health*, 94(10): 1667–71.

7 Childress, J., Meslin, E., and Shapiro, H. (eds) (2005), *Belmont Revisited: Ethical Principles for Research with Human Subjects*, Washington, DC: Georgetown University Press.

8 Childress, Meslin, and Shapiro, *Belmont Revisited*.

9 Ehni, H.-J. (2008), 'Dual Use and the Ethical Responsibility of Scientists', *Archivum Immunologiae et Therapiae Experimentalis*, 56: 147–52; Somerville, M. A. and Atlas, R. M. (2005), 'Ethics: A Weapon to Counter Bioterrorism', *Science*, Policy Forum, 25 March, p. 1881; Nixdorff, K. and Bender, W. (2002), 'Ethics of University Research, Biotechnology and Potential Military Spin-off', *Minerva*, 40: 15–35.

10 Gelsinger died during a controversial clinical trial of gene therapy. Some have argued that the principal investigator's dual role as both a researcher and a major shareholder of a private company involved with the trial was a conflict of interest that helped precipitate a number of errors.

11 Musil, R. K. (1980), 'There Must be More to Love than Death: A Conversation with Kurt Vonnegut', *The Nation*, 231(4): 128–32.

12 Jones, N. L. (2007), 'A Code of Ethics for the Life Sciences', *Science, Engineering Ethics*, 13: 25–43.

13 Miller, S. and Selgelid, M. J. (2003), 'The Ethics of Dual-use Research', in S. Miller (ed.), *Ethical and Philosophical Consideration of the Dual-use Dilemma in the Biological Sciences*, The Netherlands: Springer Sciences, NV (Chapter 3).

14 See, Preston, *The Demon in the Freezer*.

15 Guston, D. H. and Sarewitz, D. (2002), 'Real-time Technology Assessment', *Technology in Society*, 24(1–2): 93–109; Corneliussen, F. (2006), 'Adequate Regulation, a Stop-gap Measure, or Part of a Package?', *EMBO Reports*, 7: s50–4.

Chapter Nine

1 Merton, R. K. (1942), 'Science and Technology in a Democratic Order', *Journal of Legal and Political Sociology*, 1: 115–26. Reprinted as Merton, R. K. (1973), 'The Normative Structure of Science', in *The Sociology of Science. Theoretical and Empirical Investigations*, Chicago, IL: University of Chicago Press, pp. 267–78.

Bibliography

Ach, J. S. and Siep, L. (2006), *Nano-bio-ethics: Ethical Dimensions of Nanobiotechnology*, London: Transaction Publishers.

Allhoff, F. and Lin, P. (2009), *Nanotechnology and Society: Current and Emerging Ethical Issues*, The Netherlands: Springer.

Atlas, R. M. and Dando, M. (2006), 'The Dual-use Dilemma for the Life Sciences: Perspectives, Conundrums, and Global Solutions', *Biosecurity and Bioterrorism: Biodefense Strategy, Practice, and Science*, 4(3): 276–86.

Austin, J. (1832), *The Province of Jurisprudence Determined* (ed. W. Rumble), Cambridge: Cambridge University Press.

Baines, P. (2005), *Penguin by Design, A Cover Story 1935–2005*, London: Penguin Books.

Benedikt, M. (1991), *Cyberspace: First Steps*, Cambridge, MA: MIT Press.

Berne, R. W. (2006), *Nanotalk: Conversations with Scientists and Engineers about Ethics, Meaning, and Belief in the Development of Nanotechnology*, Hillsdale, NJ: Lawrence Erlbaum Associates.

Boucher, P. M. (2008), *Nanotechnology: Legal Aspects*, Boca Raton, FL: CRC Press.

Braudel, F. (1992), *Civilization and Capitalism: Vol. 2, The Wheels of Commerce* (trans. S. Reynolds), Berkeley, CA: University of California Press.

Bronowski, J. (1973), *The Ascent of Man*, Toronto, ON: Little, Brown and Company.

Cameron, N. M. de S. and Mitchell, M. E. (2007), *Nanoscale: Issues and Perspectives for the Nano Century*, Hoboken, NJ: John Wiley & Sons.

Campbell-Kelly, M. and Aspray, W. (2004), *Computer: A History of the Information Machine*, Boulder, CO: Westview Press.

Cello, J., Paul, A. V., and Wimmer, E. (2002), 'Chemical Synthesis of Poliovirus cDNA: Generation of Infectious Virus in the Absence of Natural Template', *Science*, 297(5583): 1016–18. DOI: 10.1126/science.1072266. PMID 12114528.

Cheung, S. N. S. (1982), 'Property Rights in Trade Secrets', *Economic Inquiry*, 20(1): 40.

Childress, J., Meslin, E., and Shapiro, H. (eds) (2005), *Belmont Revisited: Ethical Principles for Research with Human Subject*, Washington, DC: Georgetown University Press.

Cohen, H. W., Gould, R. M., and Sidel, V. W. (2004), 'The Pitfalls of Bioterrorism Preparedness: The Anthrax and Smallpox Experiences', *American Journal of Public Health*, 94(10): 1667–71.

Corneliussen, F. (2006), 'Adequate Regulation, a Stop-gap Measure, or Part of a Package?', *EMBO Reports*, 7: s50–4.

Coryton, J. (1855), *The Law of Letters-patent*, London: H. Sweet.

Daniels, P. and Bright, W. (1996), *The World's Writing Systems*, Oxford: Oxford University Press.

Dipert, R. (1993), *Artifacts, Art Works, and Agency*, Philadelphia, PA: Temple University Press.

Drexler, E. (1986), *Engines of Creation: The Coming Era of Nanotechnology*,
New York: Anchor Books.

Drexler, E. (1992), *Nanosystems: Molecular Machinery, Manufacturing, and
Computation*, New York: John Wiley & Sons.

Drexler, E., Forrest, D., Freitas, R. A. Jr., Hall, J. S., Jacobstein, N., McKendree, T.,
and Merkle, R. (2001), *On Physics, Fundamentals, and Nanorobots*, Institute for
Molecular Manufacturing.

Dworkin, R. (1977), *Taking Rights Seriously*, London: Duckworth.

Ehni, H.-J. (2008), 'Dual Use and the Ethical Responsibility of Scientists',
Archivum Immunologiae et Therapiae Experimentalis, 56: 147–52.

Garfinkel, M. S., Endy, D., Epstein, G. L., and Friedman, R. M. (2007), *Biosecurity
and Bioterrorism: Biodefense Strategy, Practice, and Science*, 5(4): 359–62.
DOI: 10.1089/bsp.2007.0923.

Gershenfeld, N. (2005), *FAB: The Coming Revolution on Your Desktop – From
Personal Computers to Personal Fabrication*, New York: Basic Books.

Glaberson, W. (1990), 'Love Canal: Suit Focuses on Records from 1940's',
The New York Times, 22 October.

Gladwell, M. (2008), *Outliers*, London: Little, Brown and Company.

Greenberg, D. (2007), *Science for Sale: The Perils, Rewards, and Delusions of
Campus Capitalism*, Chicago, IL: University of Chicago Press.

Guston, D. H. and Sarewitz, D. (2002), 'Real-time Technology Assessment',
Technology in Society, 24(1–2): 93–109.

Hart, H. L. A. (1958), 'Positivism and the Separation of Law and Morals',
Harvard Law Review, 71: 593 repr. in his (1983) *Essays in Jurisprudence and
Philosophy*, Oxford: Clarendon Press.

Heim, M. (1993), *The Metaphysics of Virtual Reality*, New York, NY: Oxford
University Press.

Jones, N. L. (2007), 'A Code of Ethics for the Life Sciences', *Science, Engineering
Ethics*, 13: 25–43.

Jones, R. A. L. (2008), 'Rupturing the Nanotech Rapture', *IEEE Spectrum*'s Special
Report: The Singularity, June.

Joy, B. (2000), 'Why the Future Doesn't Need Us', *Wired*, 8(4), April.

Kelsen, H. (1967), *Pure Theory of Law* (trans. M. Knight), Berkeley, CA: University
of California Press.

Klein, K. E. (2005), 'Smart Answers, "Avoiding the Inventor's Lament"', *Business
Week*, 10 November.

Koepsell, D. (2000), *The Ontology of Cyberspace: Law, Philosophy, and the
Future of Intellectual Property*, Chicago, IL: Open Court.

Koepsell, D. (2009), 'Let's Get Small: An Introduction to Transitional Issues in
Nanotech and Intellectual Property', *Nanoethics*, 3(2): 157–66.

Koepsell, D. (2009), 'On Genies and Bottles: Scientists' Moral Responsibility
and Dangerous Technology R&D', *Science and Engineering Ethics*,
16(1): 119–33 (in print). Online first/Open Access DOI: 10.1007/
s11948-009-9158-x.

Koepsell, D. (2009), *Who Owns You? The Corporate Gold Rush to Patent Your
Genes*, Wiley-Blackwell.

Koepsell, D. (2010), 'Authorship and Artifacts: Remaking IP Law for Future
Objects', *The Monist*, 93(3): 481–92.

Koepsell, D. (2011), 'Things in Themselves: Redefining Intellectual Property in the Nano-age', *Journal of Information Ethics*, 21(1).

Kot, G. (2009), *Ripped: How the Wired Generation Revolutionized Music*, New York: Scribner.

Kuhn, T. (1962), *The Structure of Scientific Revolutions*, Chicago, IL: University of Chicago Press.

Kurzweil, R. (2000), *The Age of Spiritual Machines*, London: Penguin Books.

Locke, J. (1988), *Two Treatises of Government* (ed. P. Laslett), Cambridge: Cambridge University Press.

Machlup, F. and Penrose, E. (1950), 'The Patent Controversy in the Nineteenth Century', *The Journal of Economic History*, 10(1): 1–29.

Madden, M. (2009), 'The State of Music Online: Ten Years after Napster', *Pew Internet Reports*, 15 June. Available at http://www.pewinternet.org/Reports/2009/9-The-State-of-Music-Online-Ten-Years-After-Napster/The-State-of-Music-Online-Ten-Years-After-Napster.aspx?view=all [accessed 22 September 2010].

McLuhan, M. and Fiore, Q. (1967), *The Medium is the Massage* [*sic.*], London: Penguin Books.

Menger, C. (1963), *Problems of Economics and Sociology* (trans. F. J. Nock, ed. L. Schneider), Urbana, IL: University of Illinois Press.

Merton, R. K. (1942), 'Science and Technology in a Democratic Order', *Journal of Legal and Political Sociology*, 1: 115–26. Reprinted as Merton, R. K. (1973), 'The Normative Structure of Science', in *The Sociology of Science. Theoretical and Empirical Investigations*, Chicago, IL: University of Chicago Press, pp. 267–78.

Milgram, S. (1963), 'Behavioral Study of Obedience', *Journal of Abnormal and Social Psychology*, 67: 371–8. DOI: 10.1037/h0040525.

Milgram, S. (1973), 'The Perils of Obedience', *Harper's Magazine*, December, pp. 62 77.

Miller, J. C., Serrato, R., Represas-Cardenas, J. M., and Kundahl, G. A. (2004), *The Handbook of Nanotechnology: Business, Policy, and Intellectual Property Law*, New Jersey: John Wiley & Sons.

Miller, S. and Selgelid, M. J. (2003), 'The Ethics of Dual-use Research', in S. Miller (ed.), *Ethical and Philosophical Consideration of the Dual-use Dilemma in the Biological Sciences*, The Netherlands: Springer Sciences, NV (Chapter 3).

Morton, D. L. Jr. and Gabriel, J. (2007), *Electronics: The Life Story of a Technology*, Maryland: The Johns Hopkins University Press.

Moser, P. (2005), 'How Do Patent Laws Influence Innovation? Evidence from Nineteenth-century World's Fair', *American Economic Review*, 95(4): 1214–36.

Munn, K. and Smith, B. (2008), *Applied Ontology: An Introduction*, Frankfurt/Lancaster: Ontos.

Nixdorff, K. and Bender, W. (2002), 'Ethics of University Research, Biotechnology and Potential Military Spin-off', *Minerva*, 40: 15–35.

Pelley, J. and Saner, M. (2008), 'International Approaches to the Regulatory Governance of Nanotechnology', RGI Report, School of Public Policy and Administration at Carleton University in Canada.

Peterson, C. (2001), *Rebuttal to Smalley's Assertion that Self-replicating Mechanical Nanorobots Are Simply Not Possible*, Institute for Molecular Manufacturing.

Preston, R. (2003), *The Demon in the Freezer*, New York: Fawcett.

Rawls, J. (1999), *Lectures on the History of Moral Philosophy* (ed. B. Herman), Cambridge, MA: Harvard University Press.

Rawls, J. (1999), *A Theory of Justice*, revised edn, Cambridge, MA: Harvard University Press.

Reinach, A. (1913), 'The *A Priori* Foundations of Civil Law' (trans. J. Crosby), *Aletheia*, 3: 1–142.

Roco, M. C. and Bainbridge, W. S. (2006), *Nanotechnology: Societal Implications*, Dordrecht, The Netherlands: Kluwer.

Rosen, W. (2010), *The Most Powerful Idea in the World*, New York: Random House.

Schummer, J. and Baird, D. (2006), *Nanotechnology Challenges: Implications for Philosophy, Ethics and Society*, Singapore: World Scientific Publishing.

Searle, J. R (1997), *The Construction of Social Reality*, New York: Free Press.

Seed, P. (2001), 'Taking Possession and Reading Texts: Establishing the Authority of Overseas Empires', in S. N. Katz, J. M. Murrin, and D. Greeberg (eds), *Colonial America: Essays in Politics and Social Developments*, 5th edn, New York: McGraw-Hill, pp. 22, 26.

Simons, P. and Melia, J. (2000), 'Continuants and Occurrents', *Proceedings of the Aristotelian Society*, Supplementary Volume 74: 59–75, 77–92.

Smalley, R. E. (2001), 'Of Chemistry, Love, and Nanobots', *Scientific American*, 285: 76–7.

Smith, B. D. (1995), *The Emergence of Agriculture*, Scientific American Library.

Sobel, D. (1996), *Longitude: The True Story of a Lone Genius Who Solved the Greatest Scientific Problem of His Time*, London: Penguin Books.

Somerville, M. A. and Atlas, R. M. (2005), 'Ethics: A Weapon to Counter Bioterrorism', *Science*, Policy Forum, 25 March, p. 1881.

Stephenson, N. (2000), *The Diamond Age*, New York: Bantam Books.

Taebi, B. (2010), 'Sustainable Energy and the Controversial Case of Nuclear Power', in R. Raffaelle, W. Robison, and E. Selinger (eds), *Sustainability Ethics: 5 Questions*, Copenhagen: Automatic Press.

The National Commission for the Protection of Human Subjects of Biomedical and Behavioral Research (1979), 'The Belmont Report: Ethical Principles and Guidelines for the Protection of Human Subjects of Research', 18 April.

Thurston, R. (2006), 'HP: Open Source Can be More Profitable than Proprietary'. Available at http://www.ZDNet.co.uk [accessed 29 October 2010].

Vinge, V. (1993), 'Technological Singularity' paper presented at the Vision 21 Symposium, expanding on a concept developed in his short story 'Marooned in Realtime' as well as his article 'First Word' in *OMNI* magazine, 1983.

Watt, S. (1997), 'Introduction: The Theory of Forms (Books 5–7)', Plato: *Republic*, London: Wordsworth Editions, pp. xiv–vi.

Welsh, C. A. (1948), 'Patents and Competition in the Automobile Industry', *Law and Contemporary Problems*, 13(2): The Patent System (Spring), 260–77.

Williams, H. (2009), 'Intellectual Property Rights and Innovation: Evidence from the Human Genome', 30 December, unpublished manuscript.

Available at http://www.nber.org/~heidiw/papers/5_12_10a_hlw.pdf
[accessed 22 September 2010].
Williams, T. and Singer, C. J. (1954), *A History of Technology*, vols. 1–6, Oxford:
Clarendon Press.
Wolf, A. (1950), *A History of Science, Technology, and Philosophy*, vol. I,
New York: Harper Torchbooks.

Statutes
35 US Code sec. 154(a)(2).
35 US Code sec. 1-376, The US Patent Act.
35 USC § 200–212[1], and implemented by 37 C.F.R. 401.

Cases
Parker v. Flook, 437 US 584 (1978).
Gottschalk v. Benson, 409 US 63 (1972).
Diamond v. Diehr, 450 US 175 (1981).

Index